Arboriculture

The Utility Specialist Certification Study Guide
Randall H. Miller • Geoffrey Kempter

Utility Arboriculture is intended to serve as a reference, providing general information about utility arboriculture. It is not intended to be the only program of study to obtain the ISA Certified Arborist Utility Specialist® credential. The practices and recommendations contained in this guide should be used in practice only by those properly trained, educated, and experienced in the field of arboriculture. ISA is responsible only for the educational program contained in this guide and not for the use or misuse of these ideas in specific field situations or by inexperienced or improperly trained individuals.

ISBN: 978-1-943378-01-2

Copyright © 2018 by International Society of Arboriculture
All rights reserved. Printed in the United States of America. Except as permitted under the United States Copyright Act of 1976, no part of this publication may be reproduced or distributed in any form or by any means or stored in a database or retrieval system without the prior written permission of the International Society of Arboriculture (ISA).

Production Coordinator: Kathy Ashmore
Editorial Coordinator: Marni Basic
Design and Composition: Jody A. Boles
Illustrators: Troy Courson and Beatriz Pérez González

International Society of Arboriculture
270 Peachtree St., NW
Ste. 1900
Atlanta, GA 30303
www.isa-arbor.com
permissions@isa-arbor.com

Photo on p. vi: Aaron H. Bynum; p. 2: Mike Chedester/AEP Ohio; p. 26: Geoffrey Kempter; p. 56: Justin Rastovac/Asplundh; p. 80: David Fleischner/Trees Inc.; p. 124: Geoffrey Kempter; p. 162: Mike Reynolds/Asplundh; p. 194: Ildar Sagdejev/Wikimedia Commons; p. 216: Geoffrey Kempter.

Printed by Premier Print Group, Champaign, IL
10 9 8 7 6 5
04-23/CA/1000

CONTENTS

Acknowledgments.................................v
Introduction.................................. vii

Chapter 1 Safety

Introduction...4
Safety and Utility Arboriculture.........................4
Employer Responsibilities 5
Workers 6
Behavior-Based Safety...............................8
Risks of an Arboricultural Workplace 13
Electrical Safety Standards.............................14
Electrical Safety Precautions 16
Creating a Culture of Safety 21
Summary......................................22
Chapter 1 Workbook...23

Chapter 2 Program and Personnel Management

Introduction...28
Program Planning 28
Project Management.............................32
Budgeting.......................................36
Contracting 40
Fleet Management..................................45
Personnel Management 47
Summary.......................................52
Chapter 2 Workbook...53

Chapter 3 Utility Pruning

Introduction.......................................58
Purpose of Utility Pruning 58
Utility Pruning Overview 60
Utility Pruning Objectives 66
Pruning Intervals......................71
Palm Pruning 74
Remote Forested Environments......................75
Summary..................................77
Chapter 3 Workbook...78

Chapter 4 Integrated Vegetation Management

Introduction...................................82
Setting Objectives.............................84
Evaluating the Site 90
Defining Action Thresholds.......................94
Evaluating and Selecting Control Methods 95
Implementing Control Methods.....................108
Monitoring Treatment and Quality Assurance..109
Chemicals 109
Summary......................................119
Chapter 4 Workbook..120

Chapter 5 Electrical Knowledge

Introduction 126
Electrical Fundamentals 126
Powerhouse to Electrical Outlets...................128
Electrical Hardware 133
Tree-Caused Electrical Service Interruptions ...151
Vegetation Management Regulations 154
Summary..158
Chapter 5 Workbook......................................159

Chapter 6 Storm Preparation and Response

Introduction 164
Risk Identification 165
Pre-Coordination and Preparation 172
Response 177
Incorporating Lessons Learned 190
Recognition of Employees 190
Summary..190
Chapter 6 Workbook......................................191

Chapter 7 Communications

Introduction 196
Importance of Customer Communications....197
Identifying Stakeholder Interests 199
Outreach Methods 205
Talking With Customers 207
Summary..212
Chapter 7 Workbook......................................213

Glossary 217
References...236
Answers to Workbook Questions...................251
Index..254

ACKNOWLEDGMENTS

The authors thank the reviewers and others who provided valuable input and guidance, including John Goodfellow, Dr. Jim Clark, Dr. Phillip Charlton, Craig Kelly, Jennifer Arkett, Dr. John Ball, Jim Barnhart, Anne Beard, Pedro Mendes Castro, Rick Deer, Jim Downie, Brian Fisher, Randall Gann, Dr. Edward Gilman, Lynn Grayson, Barry Grubb, Rick Johnstone, Rob Kaiser, Jean Larivière, Sharon Lilly, Tina Melton, Mike Neal, Chris O'Brien, Lisa Randel, Matthew Simons, Ruth Stein, Chuck Tally, Dr. Robert L. Tate, Jeff Treu, and Derek Vannice.

In addition, the following individuals and companies provided assistance: Mike Weidner and Troy Ross with ACRT; Joe Lentz, Koby Cutchell, Todd Hagenbuch, Jimmy Wilfong, and the staff at Arborchem; Kristin Wild, Jim Orr (retired), Dave Krause, John McNamee, and Tracy Hawks with Asplundh Tree Expert, LLC; William Porter at CNUC (retired); Paul Appelt (retired) and Ryan Brockbank at ECI; Dr. Robert Vanderhoof at PacifiCorp; Steve Tankersley at Pacific Gas and Electric (retired); Tom Prosser and Brandon Hughson at Rainbow Treecare Scientific Advancements; David Fleischner, Brent Jeffrey, Jayson Heltsley, and Jeremy Wyant with Trees, LLC; Wilber Nutter with Wright Tree Service; and Stephen R. Cieslewicz, independent consultant.

Finally, we are indebted to these individuals whose leadership and vision have profoundly influenced utility arboriculture and the contents of this volume:

Dr. Richard Abbott

Christopher B. Asplundh, Sr.

Dr. William Bramble

Dr. William Byrnes

Hyland Johns

Dr. Alex Shigo

INTRODUCTION

Vegetation has competed with the built environment for thousands of years. While few records of the earliest vegetation management efforts exist, there can be little doubt that ancient civilizations quickly learned that the same trees that provided building materials, fuel, and fodder could also play havoc when they grew in the wrong place, or failed catastrophically.

Modern utility arboriculture dates back to the 1840s when the telegraph made instantaneous communication possible across great distances. From the start, practitioners of what is now known as utility arboriculture struggled to maintain these services despite the ongoing assault of wind and weather, and a dearth of knowledge of safety and the science of tree care. Furthermore, their efforts were invariably met with misunderstandings and controversy on the part of the public.

Over the decades, the technology and science available to vegetation managers has improved. Standards, best practices, and credentialing programs for safety and for tree care have been developed and adopted by the industry; this book is written to in accordance with the latest versions of industry standards and best practices in effect at the time of publication. At the same time, the materials and equipment used to generate and distribute electricity have also evolved and been improved. As a result of these efforts, the safety and reliability of utility services has improved to the point where it is now considered to be essential by end users and government regulators.

One of the most significant challenges that continues to confront utility arborists is balancing the benefits provided by both trees and utilities. This was summarized by G.D. Blair in the preface to his classic 1940 text *Tree Clearance for Overhead Lines: A Textbook of Public Utility Forestry*:

> *A common sense basic appraisal of the factors involved (with tree-overhead line conflicts) should consider trees as nature's gift to society and the service supplied by power and communication lines as gifts of man's ingenuity. Obviously, each of these contributions is indispensable to human needs, convenience, and pleasure. To provide their greatest individual usefulness, trees must be vigorous and beautiful; overhead line service must be continuous and dependable. In this measure of quality, each is essential to the happiness of civilized people.*

It is clear that in 1940, Blair fully appreciated the "indispensable" contribution of both trees and utility services. The competition for space continues, even as the value of urban forests is better quantified, and the need for "continuous and dependable" utility service has never been greater. However, while most people appreciate the value of trees, far fewer fully understand the service reliability and safety risks posed by them. This helps explain why public resistance to utility vegetation management is so widespread. It also places utility arborists directly in the middle of these competing interests.

UTILITY VEGETATION MANAGEMENT

To meet the public's expectation of safe, reliable services, utilities direct significant resources toward keeping vegetation in check. Most of the labor is performed by specialized utility vegetation management contractors, while the management and planning of the work is divided between utilities and contract consulting firms. Together, these efforts employ tens of thousands with a variety of skills, including consultants, safety experts, managers, equipment operators, climbers, herbicide applicators, quality inspectors, and customer relations specialists. In addition, thousands more are employed in support industries, such as research, industry associations, and providers of equipment, tools, and supplies. Collectively, these disparate groups make up the utility vegetation management industry.

Electric and gas utilities maintain vast networks of transmission and distribution lines, much of which can be affected by vegetation. In general, where there are people and dwellings, whether in cities or remote rural locales, there will be utility services. Overhead electric distribution comprises the bulk of the work performed in utility vegetation management; however, maintaining compatible vegetation on electric and gas transmission corridors is also a high priority. In the process of accomplishing these tasks, face-to-face interactions with millions of utility customers and other stakeholders occur every year.

Given what is at stake, the importance of utility vegetation management cannot be overstated. There is an ongoing need to develop professionals who can safely and effectively carry out the complicated tasks of assessing, budgeting, planning, and executing the work. The efforts of those at the International Society of Arboriculture (ISA) and its affiliate, the Utility Arborist Association (UAA), including the writing of this guide, are part of an ongoing effort to raise the profile of utility arboriculture among employees, utilities, regulators, and the public.

WE ARE NOT TREE TRIMMERS

It is the opinion of the authors that the terms "tree trimming" and "tree trimmers" should be abolished from the collective lexicon of our industry. The word "trim" has many meanings—budgets, fingernails and hair can all be trimmed—but "pruning" is a scientifically-based, specialized skill requiring professional expertise. "Tree trimming" is used by others in the industry—including even a few utility executives—as a term of derision (e.g., "It's just tree trimming"), and demonstrates a lack of professional recognition. Unfortunately, our training manuals, job titles, and specifications are thoroughly contaminated with these terms, due to many decades of misunderstanding and tradition.

It should be clear that the terms "trim" and "trimmer" inadequately describe the skills required of utility arborists. Wherever "trim" and its derivations appear, they should be purged and replaced by more appropriate and professional terminology, e.g., "tree trimmers" are "utility arborists," and "tree trimming" is "utility arboriculture." Trees are not "trimmed," they are "pruned"; there is no such thing as a "tree-trimming program," it is a "vegetation management program." These more accurate terms promote utility arboriculture as a profession and will generate greater respect from the public and our industry peers.

PURPOSE OF THIS BOOK

This book provides utility arborists and foresters with the basic information required to navigate the increasingly complex requirements of working in today's vegetation management industry. It also serves as the study guide for the ISA Certified Arborist Utility Specialist® Exam. However, it should be clearly understood that not all material that is covered by the exam is included in the book. In addition, although the writers have made a sincere effort to cover a broad range of relevant information, in such a complex industry there will always be ideas and concepts that were not included but could be relevant.

Another important principle to keep in mind is that there is an inherent conflict of interest if an organization, in this case, ISA, provides both the credential and the sole source of information for that same credential. To avoid this conflict, and to remain in keeping with best practices for credentialing, ISA offers this study guide with the understanding that it provides an overview of the type of information covered on the exam, but other sources of information, including other related texts and the candidate's own experience, should also be considered.

ISA developed the ISA Certified Arborist Utility Specialist® credential in the mid-1990s at the request of the UAA. While the UAA considered the ISA Certified Arborist® program to be an excellent general arboricultural certification, members also desired a credential that covered the unique needs of utility arborists. A committee was assembled, with representatives from utilities, contractors, consulting companies, and academia. Originally, the intent was to credential field-level personnel, but after some consideration, the UAA decided to direct the focus on those who write specifications, and plan, oversee, and inspect the work. Once the details were agreed upon, the ISA Certification Board approved the ISA Certified Arborist Utility Specialist® credential, the first credential beyond ISA Certified Arborist®.

Prerequisites for obtaining the ISA Certified Arborist Utility Specialist® credential include holding the ISA Certified Arborist® credential and documented experience in utility vegetation management. Therefore, this book is written with the assumption that readers are already conversant in the basics of arboriculture and the utility industry. This book is peer reviewed, covers the latest methods and techniques, and is based on scientific research and proven best practices. It is divided into seven chapters: Safety, Program and Personnel Management, Utility Pruning, Integrated Vegetation Management, Electrical Knowledge, Storm Preparation and Response, and Communications; it also includes a comprehensive glossary of terms.

ONGOING CHALLENGES

In an effort to educate the public about the cost and risk of trees near overhead lines, many utilities promote the idea of "right tree, right place." This concept acknowledges that trees are valuable, but only when they do not have the potential to interfere with overhead utility services. However, promoters of the benefits of urban forests, armed with hard data that quantifies the value of "green infrastructure," including canopy provided by large trees, continue to challenge the idea that trees must make way for utilities. Navigating this ongoing conflict is just one of many challenges facing utility arborists.

We now have the ability to quantify the benefits provided by trees and utility services, and to provide care and monitoring to minimize risk. However, G.D. Blair's 1940 vision of vigorous and beautiful trees coexisting with continuous and dependable overhead utility services can only be realized if all stakeholders—including tree advocates, utility service providers, and government regulators—actively seek a common goal: to optimize risks and benefits, using sound science and professional expertise. Only then can utility arborists deliver the maximum value to the citizens of our communities and customers of utilities—who, after all, are one and the same. It is our hope that this guide makes a contribution to that end.

Utility Arboriculture

1
Safety

OBJECTIVES

- Clarify employer and worker safety responsibility.
- Utilize behavior-based safety principles to decrease the likelihood of safety incidents.
- Describe the potential injuries electricity can cause due to exposure to the human body.
- Practice electrical safety precautions to prevent direct or indirect contact.
- Identify the elements of a culture of safety.

KEY TERMS

- accident
- accident pyramid
- ANSI Z133
- back feed
- behavior-based principles
- close calls
- direct contact
- electric shock
- electrocution
- equipment failure
- high-reliability organization
- human error
- incident
- indirect contact
- lagging indicators
- leading indicators
- maliciousness
- minimum approach distance
- multiple causation theory
- Occupational Safety and Health Act
- Occupational Safety and Health Administration (OSHA)
- safety committee
- step potential
- touch potential
- unsafe acts
- unsafe working conditions
- willful rule violation

INTRODUCTION

Tree work has unique potential risks due to the heavy loads, work height, power equipment, and extreme weather conditions associated with the trade (Blair 1989; Ball and Vosberg 2010). Add to that the possible danger related to working in proximity to high voltage, and it becomes clear why utility arboriculture demands particular safety emphasis.

There are humanitarian, as well as economic, reasons for worker safety (Blair 1940). Heinrich et al. (1980) consider industrial safety to be a moral imperative. There is no business worth asking people to sacrifice what they value most—their health, physical ability, or lives. People work to help provide for themselves and their families and to contribute to society. Serious work injuries not only potentially degrade victims' quality of life, but also hamper their ability to work. In extreme cases, this may mean that people end up becoming dependent on the family for whom they had been providing and the community to which they had once fully contributed. So, these consequences extend far beyond the individual, reaching to family, friends, colleagues, and people in the community, who are all negatively affected by workplace safety lapses.

Safety managers often make a distinction between accidents and incidents. The term **accident** is in disfavor because it implies that outcomes are due to fate or bad luck, which is now considered to be a counterproductive message (Salmone and Pons 2007). Rather, industrial safety teaches that injuries are avoidable. Consequently, unplanned, undesirable events that could result in unintentional injuries or fatalities are better referred to as **incidents**.

From an economic perspective, incidents have a negative effect on a company's bottom line, as they generate compensation claims and medical costs. Worse, those costs account for only 20 percent of industrial injury expenditures, which increase due to higher insurance premiums, lost production, training of replacements, and other financial liabilities (Heinrich et al. 1980).

SAFETY AND UTILITY ARBORICULTURE

There are no published safety data related specifically to utility arboriculture. However, in the United States, the Bureau of Labor Statistics reports that tree work is among the highest-risk vocations in the country, along with commercial fishing and logging. The three most common events or exposure categories for fatal incidents associated with tree work are contact with objects or equipment ("struck-bys"), falls, and exposure to an unsafe environment or harmful substance. Falling limbs or trees account for the most struck-by fatalities. The majority of falls are from trees, followed by those involving aerial lifts (either due to lift failure or falls). Exposure to an unsafe environment or harmful substances is the third most frequent cause of tree worker death. Within this event or exposure category, the deaths are almost entirely due to "exposure to electricity"—electrocution. (Ball and Vosberg 2010).

Incidents can be attributed to **unsafe acts** and **unsafe working conditions**. Unsafe acts include **human error**, **willful rule violation**, and **maliciousness**. **Equipment failure** is an example of an unsafe condition. Managers can plan for and prevent most unsafe conditions. Equipment failure should be minimized through regular inspection and maintenance. Providing workers with the appropriate, mandatory personal protective equipment; including helmets, eye and hearing protection and cut-resistant chaps; reduces injury severity resulting from incidents. Unsafe acts, such as willful rule violation and maliciousness, cannot be tolerated and should be subject to swift and decisive punishment. Human error is more difficult to manage because under normal circumstances, most people average five errors an hour and can only concentrate on two or three things simultaneously. However, people often make errors when they are assigned tasks for which they are unsuited or inadequately trained (Petersen 2001). Agnew and Daniels (2010) suggest a "Can't do/Won't do" test to determine whether an incident is attributable to deficient training or an unsafe workplace condition. If an employee can't do what was required even if their life depended on it, it is a training problem. If they could do it, but didn't, it is a behavior or motivational problem.

EMPLOYER RESPONSIBILITIES

In addition to industry's moral responsibility and financial self-interest in preventing incidents, in many countries employers also have legal requirements to provide a safe workplace. For example, in the UK, the *Health and Safety Act of 1974* requires that, as far as reasonably possible, employers ensure the health, safety, and welfare of their employees. In the United States, enforcement of safety work rules resides with the **Occupational Safety and Health Administration (OSHA)**. The "General Duty" clause of the **Occupational Safety and Health Act** requires employers to provide their employees with working conditions that are free of known dangers (OSHA 2016). Employers must provide personal protective equipment free of charge and inform employees of their rights and duties under their [the employer's] safety and health program (Van Soelen 2006). For arboricultural operations in the United States, employers must implement a structured safety program following **ANSI Z133** (ANSI 2017a) and other standards (see the *Electrical Safety Standards* section).

Heinrich et al. (1980) complain that management has not applied to incident prevention the same communication, responsibility, authority, and accountability tools that have historically proven effective to control quality, productivity, and cost. Rather, they contend that management flounders along with safety posters, contests, and safety committees, and then punishes people who are involved in incidents. They submit that the same management strategies that drive improved quality and productivity can also effectively promote safety, and it is a mistake to neglect to bring those strategies to bear on incident prevention. In that regard, Petersen (2001) advises that safety programs should have at their foundation an analysis of corporate requirements, employee needs, and the demands of the job.

Petersen (1997) identifies five ways employers create unsafe workplaces:

- Provide employees with minimal control over their work environment
- Expect employees to be passive, dependent, and subordinate
- Demand that employees use a short-term time perspective
- Induce employees to perfect and value frequent use of superficial abilities
- Insist on high production in the counterproductive work environment created by the preceding bulleted points (leading to psychological failure)

Employers often fail to develop a culture of safety because they adopt ineffective management styles. Petersen (2001) suggests that corporate safety programs should reduce dependence on common unsafe work rules and incorporate job enrichment, participation, and employee-centered leadership.

Agnew and Daniels (2010) identify seven safety practices that waste time and money:

1. Focusing on lagging indicators (concentrating on results rather than what employees are actually doing)
2. Injury-based incentive programs
3. Awareness training (instead, determine whether the problem is actually a safety issue with the can't do/won't do test)
4. Reliance on safety signage
5. Punishing those who make safety errors
6. Misunderstanding "near misses" (every deviation from normal should be reported and investigated, not just close calls)
7. Thinking that checklists will change behavior without follow-up to make sure they actually do so

Safety Committees

Companies dedicated to workplace safety should have organized **safety committees**, and they are required in many states in the United States. For example, the state of Oregon mandates that companies with more than ten employees have safety committees, although smaller firms are only required to hold safety meetings (Oregon OSHA 2016). Safety committee members and champions should be drawn from every level in a company, from crews to management. Employers have found that in many cases, the workforce imposes stricter safety policy enforcement than management, because employees want responsibility, best understand circumstances involving work practices, and are willing to make tough decisions (Nutter 2012).

The Maine Municipal Association (2011) recommends the following steps for organizing safety committees:

- Determine the structure of the committee
- Agree on the optimal size of the committee (five to ten is recommended)
- Determine who will serve on the committee (by assignment, request for volunteers, election, or other means)
- Establish criteria for rotating members on and off the committee
- Involve top management
- Review training and safety education material
- Identify high-risk tasks
- Develop a clear mission statement
- Establish an agenda protocol for the meetings
- Keep minutes and communicate outcomes throughout the organization
- Review the progress of the committee at least annually
- Develop safety procedures

The safety committee should meet at least monthly and set both short-term (one to six months) and long-term (greater than six months) objectives. The goals should be SMART (see Chapter 2). Safety committees must have the support of members of top management, who should attend at least periodically. The committee should focus on safety and not become a forum to air complaints or grievances (Maine Municipal Association 2011).

Safety committees should utilize a number of tactics (Van Soelen 2006; Society for Human Resource Management 2013):

- Develop safe work practices
- Craft written programs
- Lead safety training
- Assist in incident investigations
- Review incident investigation reports
- Monitor status of action plans from safety meetings and incident investigations
- Initiate problem-solving teams to address priority safety issues
- Participate in regular workplace condition inspections
- Participate in regular internal safety program assessments
- Promote employees' interests in health and safety issues
- Provide a setting in which labor and management can discuss health and safety issues and collaborate on solutions

WORKERS

Workers are the center of safety programs, and play a critical role in developing and maintaining a safe work environment. They have both rights and responsibilities as part of safety programs.

Worker Rights

In the United States, workers have a series of rights granted by the Occupational Safety and Health Act of 1970 (OSHA 2016). Workers have a right to:

- File a confidential complaint that initiates an OSHA workplace inspection.
- Receive information and training about hazards, methods to prevent harm, and OSHA standards that apply to their workplace. The

Achieving Compliance with Laws and Company Policies

- All employees must "buy in" to the safety process. Everyone, from the newest employee to the president, has to see the value in safety programs.
- Safety training must be more than just "check the box" compliance. Managers must show employees that they care and are not just trying to gain compliance with rules or regulations.
- Appropriate time must be allotted for safety training.
- Communication has to be open in both directions: from the top down *and* from the bottom up. Leadership should create the vision in which beliefs and behaviors are formed in order to define a culture of safety. Beliefs and behaviors cannot be compartmentalized.
- Employees must feel comfortable when asking questions and not fear they may be perceived as rebels for doing so.
- Training cannot be a "once and done" concept. Refresher training and follow-up are critical. Training for fluency should be the objective of any training program (Agnew and Daniels 2010).
- Sometimes experience can become a disadvantage if bad habits are tolerated or complacency is allowed to permeate a program.
- There must be ongoing job behavior observations and evaluations, and small irregularities should be corrected before they become habits.
- Safety must be implemented and enforced consistently.
- There must be a clearly defined accountability for actions at all levels.
- Training has to be individualized to suit the needs of the workforce. There is no one-size-fits-all program.
- Employees need to be empowered and feel like part of the process and not just the result of the process.
- Senior management must be involved and show their support of the process by setting the example with their own behaviors.
- Employees must understand that production does not conflict with safety. The safest crews are often the most productive crews.
- Management and customers must view safety training as an investment and not a cost.

(From Jacobs 2012)

---

training must be done in a way that workers can understand.
- Review the employer's records of their work-related injuries and illnesses.
- Obtain copies of their workplace medical records.
- Receive copies of the results from workplace monitoring and tests.
- Participate in an OSHA inspection and speak in private with an inspector.
- File an OSHA complaint if their employer has subjected them to retaliation or discrimination because they [the worker] requested an inspection or invoked any of their rights under the Occupational Safety and Health Act.

Worker Responsibilities

Each worker is responsible for their own safety on the work site (ANSI 2017a). They have to understand and comply with corporate work standards and rules, governmental regulations (OSHA in the United States) and other requirements, follow their supervisor's instructions, and participate in training (Van Soelen 2006). The State Compensation Insurance Fund (n.d.) asserts that workers have the following responsibilities:

- Be familiar with governmental requirements that regulate the industry
- Attend all required safety training
- Never operate equipment unless properly trained
- Read and understand the safety data sheets and know the hazards and safe work practices
- Be personally responsible. Act professionally and concentrate on personal safety and that of others. Never indulge in horseplay. Serve as a good role model, maintain a clean work area and wear personal protective equipment
- Communicate with supervisors about safety. Share suggestions to make processes or machinery safer. Notify supervisors of malfunctioning equipment, unsafe behavior, and hazardous conditions
- Report all incidents and close calls
- Report job-related injuries or illnesses

Worker Safety Tools

The Tennessee Valley Authority (2011) has identified a series of tools to improve human performance, including:

- **Job briefing.** Job briefings are used to formally review site conditions and job procedures, and to determine whether or not the work can be completed safely. Changing conditions on a job site may require an additional job briefing. They are covered later in the chapter.
- **Two-minute rule.** The two-minute rule augments the job briefing. Workers should take two minutes before starting a job to check for potential unsafe conditions, and talk over what they find with coworkers to eliminate hazards or develop contingencies before proceeding with a task.
- **Active listening.** In face-to-face communication, a "sender" wishing to communicate uses the name of the person with whom she or he is trying to communicate and states their message. The "receiver" paraphrases the message to indicate they understand the message, and if the message was properly understood, the "sender" acknowledges it was (see Chapter 7).
- **Procedure use and adherence.** Procedures ensure safe, correct, and consistent performance and must be strictly followed.
- **Self-checking S.T.A.R.** Self checking helps prevent errors, particularly during skill-based tasks like utility line clearing. The acronym has four components:

    Stop: Focus on the task. Eliminate distractions.
    Think: Concentrate on the action about to be performed. Anticipate expected results and formulate contingencies for unexpected results.
    Act: Concentrate on correct performance.
    Review: Were expected results achieved?

- **Stop when unsure.** If something doesn't seem right, stop. Talk through the situation with a coworker or supervisor.
- **Post-job review.** Look for what could be improved or what went well that can be used on future jobs.

BEHAVIOR-BASED SAFETY

Industrial safety was pioneered in the 1930s by Herbert Heinrich, who initiated research that resulted in **behavior-based principles** of accident prevention (Heinrich's term) that are still in use (Petersen 2001). Heinrich found that 88 percent of all industrial accidents are caused by people committing unsafe acts, with the remainder attributable to unsafe working conditions. Heinrich reached the commonsense conclusion that the more frequently a labor force works unsafely, the greater the chance they will have an accident. He determined that for every serious injury or fatality, there are scores of minor injuries, hundreds of **close calls,** and thousands of unsafe acts (Heinrich et al. 1980). Heinrich concluded that the best way to prevent serious or fatal accidents is to eliminate unsafe acts and conditions. He illustrated the concept with an **accident pyramid**, in an example comprising untold thousands of unsafe acts or conditions at the base of the pyramid, which led to 300 close calls, 29 minor injuries, and ultimately one major injury or fatality at the top (Heinrich et al. 1980). Jacobs (2012) considers accepting unsafe acts as contributing to a counterproductive culture characterized by what he calls normalization of deviance.

While Heinrich's work still has validity, thinking has advanced since the 1930s. Heinrich's precise ratio of one major accident for every 29 minor accidents and 300 close calls should not be viewed as representative of all industries, but merely an example to illustrate the broader concept. Terminology has changed to *incidents* rather than *accidents*—for example, to remove the notion that fate can be a significant factor. Further, Heinrich's overwhelming focus on workers' contribution to incidents and the ratio of each layer of the pyramid are also outdated. Heinrich's work was based on insurance data that was not representative of all industries, and the contribution of unsafe conditions in incidents may be, and most likely is, higher. So there is a need for companies to examine their role in prevention rather than assume it is mostly workers' faults.

McClenahan (2012) has advanced Heinrich's theories, describing the unsafe acts and close calls at the bottom of the pyramid as a safety program's **leading indicators**, and serious injuries or fatalities as **lagging indicators**. McClenahan modified Heinrich's pyramid, putting prevention-based systems at the base, classifying close calls, workplace errors, and first-aid cases under USD $100 as leading indicators, OSHA recordables under USD $500 as transition cases, and lost time injuries to be lagging indicators (Figure 1.1). While he endorses incident investigation to obtain lessons

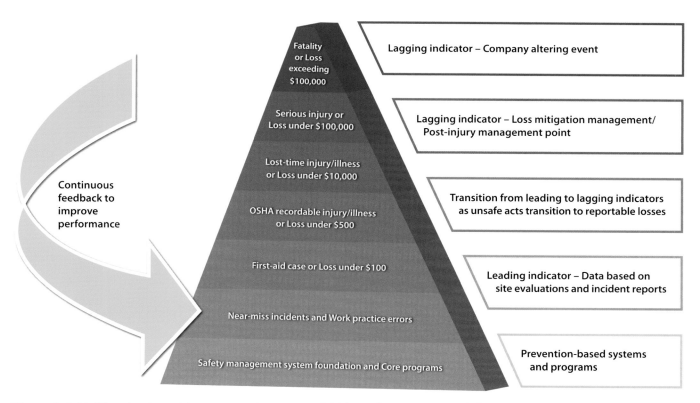

Figure 1.1 McClenahan's accident prevention pyramid. Values shown are U.S. dollars.
(Adapted, by permission, from J. McClenahan, Risk Management: Utilization of leading indicators in the continuous improvement cycle. TCI Magazine, November 2012. Based on an original diagram by H.W. Heinrich, Industrial Accident Prevention 1950)

learned from lagging indicators, he agrees with Heinrich that it is far more effective for safety programs to prevent incidents in the first place and recommends proactively using lessons learned from leading indicators to do so.

Over-Reliance on Punishment

Critics of behavioral-based safety argue that Heinrich's theory has been abused by some who use it as justification to over-rely on punishment to reduce unsafe acts (Petersen 1997). Over-reliance on punishment can be counterproductive because it carries the implication that bad things happen only to bad people, and those bad people deserve punishment. While it might be satisfying to management to blame workers for choosing to work unsafely, using punitive action as a primary safety driver is a shortcut that often undermines morale and creates a mistrustful work environment. It can compromise safety programs by discouraging employers from coming forward to report close calls out of fear of reprisal. As a consequence, retaliatory programs often lose the opportunity to fully benefit from lessons learned, including identifying latent circumstances and conditions that can lead to injuries (Reason 2000; Petersen 2000). Finally, over-reliance on punishment diverts employers from their responsibility to prevent unsafe conditions. Managers should ask themselves whether punishment has led to a desired outcome in the past, or if it is ever likely to do so, and whether or not the side effects of punishment have been counterproductive and if a positive outcome could be achieved without it (Agnew and Daniels 2010).

Agnew and Daniels (2010) argue that discipline is often misapplied as punishment rather than positive direction. They advise that the root of the word *discipline* is *disciple*. So, the objective of discipline should be to create employees who are willing followers who self-manage their behavior even if no one is watching. That can't be done with punitive measures.

Multiple Causation Theory

Multiple causation theory is a refinement of behavior-based safety theory. As described by Petersen (2001), this philosophy maintains that Heinrich's theory oversimplifies reality. Petersen considers workplace injuries to be caused by *a* number (rather than *the* number) of contributing factors and causes, which randomly interact. Prominent among these factors and causes is exposure to serious conditions. Rather than simply reducing the overall frequency of unsafe acts, Petersen reasons that safety managers should direct their attention to those circumstances that are most likely to result in severe injuries. For example, debilitating injuries are far more likely to occur to workers at height in proximity to high voltage lines than to those completing production reports. Petersen concludes that it's a distraction for safety managers to spend their valuable time eliminating unsafe acts on benign activities, when their energy would be better spent addressing behavior and conditions on potentially more threatening tasks.

Petersen (2001) maintains that severe injuries are most likely to occur under the following circumstances:

- Unusual, nonroutine work
- Nonproduction activities
- Work associated with sources of high energy (e.g., electricity, steam, compressed gas and flammable liquids, height)
- Some construction activities

From Petersen's perspective, focusing on these conditions is a more effective way of reducing severe injuries than simply reducing all cases of unsafe acts. The idea isn't that Heinrich is wrong, just that time is limited, and the pyramid concept is more effective if intervention is focused on addressing circumstances most likely to result in serious injuries or fatalities.

McClenahan (2012) advances the multiple causation theory by prioritizing risk factors through a risk assessment matrix. The matrix categorizes unsafe acts on the basis of the probability that they could cause an incident and the likely severity of the event's consequences (Figure 1.2). The approach can be used to evaluate where managers should focus their energies. For example, a frequently occurring unsafe act with catastrophic potential consequences (serious injury or death)

Ten Axioms of Industrial Safety

1. Injuries result from a series of factors, ending in an accident. Accidents (Heinrich's word) are consistently caused or permitted directly by an unsafe act, a mechanical or physical hazard, or both.
2. Unsafe acts of people are responsible for the overwhelming majority of incidents.
3. A person who suffers a disabling injury caused by unsafe acts has an average of over 300 close calls from serious injury due to the same unsafe act or mechanical hazard that resulted in the injury.
4. The extent of injuries is largely due to luck, but the incident is almost always preventable.
5. Four basic motives or reasons for unsafe acts (improper attitude, lack of knowledge or skill, physical unsuitability, and improper mechanical or physical environment) offer a guide to selecting effective preventative measures.
6. Four basic methods can be applied to accident prevention: engineering revision, persuasion and appeal, personnel adjustment, and discipline.
7. The most valuable methods of accident prevention should be consistent with the management techniques used to control quality, cost, and production.
8. Management has the best opportunity and ability to initiate the work of prevention, so it should assume that responsibility.
9. The supervisor or foreperson's influence is central to industrial accident prevention.
10. A humanitarian incentive for preventing injury is driven by two economic factors:
    a. Safety is efficient and unsafe acts are inefficient. The same lack of detail that leads to unsafe acts also leads to lapses in quality and productivity.
    b. The direct cost of industrial injuries for compensation claims and medical treatment is only 20 percent of the total costs that employers must pay.

(Heinrich et al. 1980)

would carry extreme risk and should draw close attention. On the other hand, there are probably other areas to apply efforts than the isolated acts with negligible severity. That is not to say those unsafe acts should be ignored, just not emphasized. McClenahan advises that the technique provides a systematic approach to establishing feedback loops, trend analyses, and resource allocations. He also counsels companies to maintain the criteria that are already best for their company culture, rather than waste time focusing on areas of known high compliance.

McClenahan (2012) advises using job behavior observations and perception surveys to collect leading indicators. He asserts that successfully applying this information carries the following benefits:

- Targeted training programs that address "real" issues within an organization
- Training that can advance beyond compliance only-based required programs
- Training that can be geared toward operational efficiency and attainment of valuable results
- Data that can be combined with lagging-indicator data to strengthen employee development programs
- Provision of an early-warning system
- Provision of metrics for employee performance beyond the dollars and cents of a job

Probability of Loss	Severity of Loss			
	Negligible	Marginal	Serious	Catastrophic
Improbable	Low	Low	Moderate	High
Occasional	Low	Moderate	High	Extreme
Probable	Low	Moderate	High	Extreme
Frequent	Moderate	High	Extreme	Extreme

Figure 1.2 Generic risk matrix. *(Adapted, by permission, from J. McClenahan, Risk Management: Utilization of leading indicators in the continuous improvement cycle. TCI Magazine, November 2012)*

High-Reliability Organizations

Many of the concepts of behavior-based and multiple causation theories have been successfully applied to what Weik and Sutcliffe (2007) call **high-reliability organizations**. To them, high reliability organizations, such as aircraft carriers and nuclear power plants, have excellent safety records because their consequences of failure are potentially extreme. High-reliability organizations take a system approach to safety. They recognize that humans are imperfect, and that errors have to be considered in the context of the system in which they occurred. They are characterized by involving everyone in the organization to focus on the possibility of failure, act proactively to prevent setbacks, and rehearse protocol for recovering from them. They generalize failure in an effort to establish a culture of safety (Reason 2000). Their techniques are applicable to everyone in the organization.

High-reliability organizations are characterized by five principles:

- Preoccupation with failure
- Reluctance to simplify
- Sensitivity to operations
- Commitment to resilience
- Deference to expertise

Preoccupation with Failure

A preoccupation with failure means that safety lapses are treated as an indication that something is wrong, and there are potentially serious consequences. It emphasizes reporting close calls to the extent that those who come forward with details on even minor issues (those that McClenahan [2012] would call leading indicators) are rewarded. High reliability organizations are also mindful that success can lead to complacency, automatic processing, and a temptation to reduce margins of error, which can cause a safety program to regress.

Reluctance to Simplify

High reliability organizations are reluctant to accept simple explanations for close calls or minor incidents because they can lead to lost opportunities to avoid serious incidents. Consequently, they emphasize constant vigilance and elaborate on close calls. That doesn't mean there is never a simple explanation for an irregularity; it is just that simplification should be done slowly, mindfully, and reluctantly to ensure there isn't more to a situation than originally meets the eye (Weik and Sutcliffe 2007).

Sensitivity to Operations

High reliability organizations are sensitive to operations because that is where work is accomplished. To be successful, management has to pay attention to what is being done, rather than what is supposed to be done. It must be aware that decisions made without considering practical consequences may be counterproductive (Weik and Sutcliffe 2007).

Concentrating on operations gives context to a safety program and builds effectiveness.

Commitment to Resilience

Commitment to resilience promotes an awareness of past errors and dedication to correct them before they reoccur and cause harm. High reliability organizations are proactive and prevent future incidents by not only requiring close call reporting, but also by applying the lessons learned from close calls to prevent future difficulties.

Deference to Expertise

Deference to expertise means looking for answers from those who have actual knowledge that can be brought to bear on a desired outcome. That means going beyond positions of authority for advice. High reliability organizations value leaders who ask for help in problem solving and involve the workforce in safety decision making (Weik and Sutcliffe 2007).

High-Performance Safety Culture

Agnew and Daniels (2010) describe organizations with high-performing cultures as having effective leaders who link people, shifts, and departments to simultaneously achieve high levels of productivity, quality, and safety. Effective leaders establish a culture where employees have developed a level of unconscious competence in every aspect of their work. They argue that focusing on high-impact safe behaviors at all levels facilitates fluency in delivering safe, productive, quality work.

RISKS OF AN ARBORICULTURAL WORKPLACE

Most arboricultural medical emergencies are due to trauma. Trauma is caused by outside forces exceeding the body's limits or tolerance. Struck-bys, falls, and electrical shock are the most common sources of arboricultural workplace trauma. Electrical injuries are covered in a separate section.

Both falls and being struck by a falling branch or tree involve height. The higher the height, the greater the potential threat of trauma. There are two kinds of energy, kinetic and potential. Potential energy is stored energy, ready to be released. Gravitational energy is a type of potential energy that involves the weight and height of an object. A large branch high in a tree has more potential energy than a small branch low in the canopy (Ball and Johnson 2015).

Falling objects have kinetic energy. The amount is dependent on mass and velocity, which for the sake of this discussion translates to weight and speed. Kinetic energy (KE) is expressed in the following equation:

$$KE = (weight / 2) \times speed^2$$

That means doubling the weight doubles the energy, but doubling the speed quadruples it. So, as important as weight is, height is paramount. Ball and Johnson (2015) illustrate the concept with an example of a branch section, 6 inches (15 cm) in both diameter and length, which weighs roughly 5 pounds (2.25 kg). If that branch falls 1 foot (30 cm) before striking a groundworker, it is only moving at 5 mph (8 km/h). If that same branch section falls 40 feet (about 12 m), it builds to 34 mph (55 km/h) and produces 236 foot-pounds (320 Nm) of force. To put that into perspective, 35 to 50 foot-pounds (roughly 50 to 70 Nm) are enough to inflict skull fractures (Ball and Johnson 2015).

It turns out that 40 feet (12 m) is a death threshold, where even small objects can kill. Consider that helmets are designed to withstand 36 foot-pounds (49 Nm) of force. That is roughly an 8-pound (3.5 kg) object falling 4.6 feet (1.4 m), which isn't much comfort when in tree work, where much heavier objects can fall from far more than 40 feet (12 m). This is why establishing work zones and using a call and response technique during pruning and removal operations is so important (Ball and Johnson 2015).

A falling tree or tree part usually cause blunt trauma when it strikes a worker. Blunt trauma does not break the skin, but it can still cause serious internal injury and bleeding. Crew members should be aware that anyone who has been struck by a branch that has fallen more than a few feet (about a half meter), or anyone who has fallen more than 15 feet (5 m), might be suffering from invisible, serious injury and could require medical attention. Penetrating trauma punctures skin and can cause intensive soft tissue and internal organ damage. Penetrating trauma is often caused by saws, chippers, stumpers, or branches falling end-first into a victim (Ball and Johnson 2015). Electrical shock creates penetrating trauma, and it warrants particular attention.

Electrical Injuries

Electric shock is the result of the passage of electrical current through the human body. **Electrocution** is death from electrical shock. In contrast to the powerful electric loads in high voltage lines, the human body runs on a delicate electrical system, whereby minute electrical impulses fire among the synapses of nerve cells to control muscle movement (both voluntary and involuntary). It doesn't take much current to disrupt or overwhelm that delicate electrical system. For example, contact with only 16 milliamps (0.016 amps) can disrupt a person's nervous system to the point where they can't release their grasp. Twenty milliamps (0.020 amps) can cause respiratory paralysis. That means someone contacting only 20 milliamps may not be able to let go of an electrical source and could suffocate unless the electric connection is broken in time. One-hundred milliamps (0.10 amps) can trigger ventricular fibrillation, leading to death. Two amps may cause cardiac arrest and internal organ damage, and 15 to 20 amps will likely kill outright (Table 1.1). However, someone contacting high voltage can create a fault current that builds to thousands of amps. Little wonder most people are killed outright by high-voltage contact.

Survivors often suffer permanent injuries. Human skin is a poor conductor of electricity, and its high resistance generates heat energy that is proportional to the square of the voltage (Ball and Johnson 2015). Surface temperatures at the point of contact with high voltage can reach 1,000°C (1,832°F), enough to ignite clothing and hair. Resulting burns create the threat of overwhelming infection. The danger of infection often lingers until multiple rounds of skin grafts are finally successful (Miller 2002). Inside the body, high voltage is capable of creating muscle spasms sufficiently intense to fracture bones. High voltage contact can cause severe nerve damage, potentially resulting in comas, seizures, paralysis, or other neuropathological results (Miller 2002). Bones are also resistant to electrical conduction, so the heat generated as high voltage passes through the skeletal system and can create extensive internal burns. Muscles are burned from the inside out like a hotdog in a microwave oven. In 19 percent of electrical contact cases, nerve and tissue are damaged to the point that that doctors have to amputate an affected limb, and many times multiple limbs (Ball and Vosberg 2010).

Electrical contact survivors often spend weeks or months in intensive care in critical burn units. Their injuries are excruciating. Patients may have to be put into induced comas or treated with morphine to dull their misery. Moreover, even after they have stabilized, their ordeal is not over. Many sufferers have to endure months or years of painful physical therapy as they train damaged nerves and muscles to function again and learn to live with life-long deformities and disabilities.

Table 1.1 Possible consequences of contact with various amperages.

Amperage	Possible consequence of electrical contact
0.016 amps	Motion paralysis
0.020 amps	Respiratory paralysis
0.10 amps	Ventricular fibrillation
2 amps	Cardiac arrest and internal organ damage
15–20 amps	Fatality

ELECTRICAL SAFETY STANDARDS

An excellent example of electrical safety standards is the *Z133 American National Standard for Arboricultural Operations – Safety Requirements* (ANSI 2017a). The standard has been recommended for use in Canada (Thiesen 2017) and Asia (Eckert 2017). Safe Work Australia (2016) provides no minimum approach distances, referring arborists to their state or territory electrical supply authority or electricity regulator for details on the extent of work zones, approach distances, and specific requirements that apply to working in the vicinity of power lines. UK safety standards are widely used in Commonwealth countries. The requirements discourage working near energized conductors (Eckert 2017).

Remember that employers are responsible for providing a workplace free of known hazards. One way to exercise that responsibility is for management to study, know, and implement electrical safety standards for arboricultural operations. Two important examples in the United States are OSHA 1910.269 and ANSI Z133.

ANSI Z133 is the definitive safety standard and the most current criteria in the U.S. for arborists and other workers engaged in arboricultural operations. It has the force of law because it is a document an OSHA compliance officer will often reference when citing a safety violation for arboricultural operations in the United States.

ANSI Z133 defines electrical hazard as existing any time a worker, tool, tree, or any other conductive object is closer than 10 feet (3.05 m) from an energized overhead conductor rated to 50 kV or less. The area of electric hazard expands 0.4 inch (10.2 mm) for each kilovolt above 50 kV, ultimately out to 35 feet (10.7 m) for lines rated between 785 to 800 kV (Table 1.2). In effect, **minimum approach distances** serve as a barrier between workers and electric hazards.

The 2017 revision of ANSI Z133 provides two classifications of arborists who work inside the 10-foot (3.05 m) minimum approach distances—incidental line-clearance arborists and qualified line-clearance arborists. Incidental line clearance applies when an electric hazard exists but work is not being done for the utility. Qualified line clearance applies to being in proximity to electric hazard on behalf of the utility.

ANSI Z133 requires incidental line-clearance arborists to be trained on safety-related work practices against the voltage level to which they are exposed. They are required to maintain slightly more than the minimum approach distances in Table 1.4 up to 72kV, at which point the distances

Table 1.2 Minimum approach distances to energized conductors for arborists not qualified by training and experience to work within 10 feet (3.05 m) of electrical conductors.

Nominal voltage (phase-to-phase)*	Minimum approach distance (MAD)	
kV	ft-in	m
50.0 and less	10-00	3.048
50.1 to 72.5	11-00	3.353
72.6 to 121.0	12-08	3.861
138.0 to 145.0	13-04	4.064
161.0 to 169.0	14-00	4.267
230.0 to 242.0	16-08	5.080
345.0 to 362.0	20-08	6.299
500.0 to 550.0	26-08	8.128
785.0 to 800.0	35-00	10.668

*Exceeds phase-to-ground per 29 CFR 1910.333
Reprinted, by permission, from American National Standards Institute 2017a.

become identical to those in Table 1.2. Before incidental work is done inside of the minimum approach distances (shown in Table 1.3) without insulating tools, overhead lines must be de-energized, and it is the responsibility of the system operator to make the work area safe (ANSI 2017a). Minimum approaches that pertain to incidental line-clearance arborists, inside of which no one may work, are in Table 1.3.

OSHA 1910.269 is the Occupational Safety and Health Administration's vertical standard that pertains to the generation, transmission, and distribution of electricity. A section of OSHA 1910.269 requires that anyone performing tree work in proximity to electrical hazards must be qualified, and their training has to be documented. However, this requirement is limited to qualified line-clearance arborists (or trainees under the supervision of qualified line-clearance arborists). Annex B of ANSI Z133 provides safety training recommendations to qualify workers as line-clearance arborists. Thus, OSHA 1910.269 and ANSI Z133 operate in tandem in the U.S. to direct rigorous training requirements

Table 1.3 Approach distances for incidental line clearance—alternating current.

Voltage range (phase-to-phase)*	Minimum approach distance (MAD)	
kV	ft-in	m
0.300 and less	Avoid contact	Avoid contact
0.301 to 0.750	1-06	0.457
0.751 to 5.0	2-09	0.838
5.1 to 15.0	2-10	0.864
15.1 to 36.0	3-04	1.016
36.1 to 46.0	3-08	1.118
46.1 to 72.5	4-04	1.321
72.6 to 121.0	12-08	3.861
138.0 to 145.0	13-04	4.064
161.0 to 169.0	14-00	4.268
230.0 to 242.0	16-08	5.080
345.0 to 362.0	20-08	6.300
500.0 to 550.0	26-08	8.128
785.0 to 800.0	35-00	10.668

*Exceeds Table S-5 29 CFR 1910.333.
Reprinted, by permission, from American National Standards Institute 2017a.

for anyone working within ANSI Z133's minimum approach distances on behalf of utilities. Minimum approaches that pertain to qualified line-clearance arborists are shown in Table 1.4.

Direct contact occurs when someone touches an energized fixture. For instance, a person climbing a tree and brushing up against a power line may be injured, maimed, or killed due to direct contact with high voltage. On the other hand, **indirect contact** could result when a person touches a conductive object in contact with an energized fixture. Arborists need to be aware that trees, people, various tools, fences, telecom wires, the ground, and many other objects are conductive.

Touch potential is the difference in voltage between two objects someone simultaneously contacts. Remember that that can be an energized object and another conductive, grounded object or multiple contacts with a single energized object (such as two hands, a hand and foot, or any combination of body parts). Electricity will use the body to bridge the touch potential, possibly resulting in serious injury or death.

Step potential is the difference in voltage on the ground between two body parts near where electricity goes to ground (a ground fault). As electricity goes to ground, it dissipates in concentric "ripples" similar to those formed when a rock falls in calm water. As electricity dissipates, a voltage difference builds between one "ripple" and another. Step potential develops when a person near a ground fault straddles the voltage difference with two parts of their body (such as two feet, or a hand and a foot or another body part). In such cases, electricity may jump from higher to lower voltage through a person's body. Consequently, a tree worker (or anyone else) who is merely standing near a ground fault may be injured or electrocuted by step potential, even if not touching an energized object, like a tree.

ANSI Z133 has other safety provisions and recommendations that cover more than electric hazard. One requirement is a pre-work job briefing that reviews potential hazards associated with the project, including procedures, special precautions, personal protective equipment, and work assignments, as well as electric hazards and other issues (Figure 1.3). Blair (1989) promotes the use of checklists, noting that aviators successfully use pre-flight checklists (Figure 1.4). Agnew and Daniels (2010) caution that checklists are insufficient to ensure safety unless measures are taken to ensure that they actually drive the behavior they are designed to promote.

ELECTRICAL SAFETY PRECAUTIONS

The severity of electrical injuries illustrates why the concept of active incident prevention is so important because by the time someone has contacted high voltage, it is nearly always too late. The only way to ensure safety around high voltage is to respect it and develop a safety culture based on experience and the provisions of ANSI Z133 to plan carefully to prevent direct or indirect contact.

Table 1.4 Minimum approach distances (MAD) from energized conductors for qualified line-clearance arborists and qualified line-clearance arborist trainees.

Voltage Range (phase-to-phase)* (kV)	Altitude correction factor sea level to 5,000 ft (0–1,524 m)* Phase-to-ground		Altitude correction factor 5,000 to 10,000 ft (1,524–3,048 m)* Phase-to-ground		Altitude correction factor 10,000 to 14,000 ft (3,048–4,267 m)* Phase-to-ground	
	ft-in	m	ft-in	m	ft-in	m
0.050 to 0.300	Avoid contact		Avoid contact		Avoid contact	
0.301 to 0.750	1-02	0.356	1-04	0.407	1-06	0.458
0.751 to 5.0	2-03	0.686	2-06	0.762	2-09	0.839
5.1 to 15.0	2-03	0.686	2-07	0.788	2-10	0.864
15.1 to 36.0	2-08	0.813	3-01	0.940	3-04	1.016
36.1 to 46.0	2-11	0.889	3-04	1.016	3-08	1.118
46.1 to 72.5	3-06	1.067	4-00	1.220	4-04	1.321
72.6 to 121.0	3-11	1.194	4-06	1.372	4-10	1.474
121.1 to 145.0	4-06	1.372	5-02	1.575	5-07	1.702
145.1 to 169.0	5-01	1.550	5-09	1.753	6-03	1.905
169.1 to 242.0	7-00	2.134	7-11	2.413	8-07	2.617
242.1 to 362.0	11-09	3.582	13-06	4.115	14-07	4.445
362.1 to 420.0	14-08	4.471	16-09	5.106	18-02	5.538
420.1 to 550.0	17-06	5.334	20-00	6.096	21-08	6.604
550.1 to 800.0	23-09	7.239	27-02	8.281	29-05	8.967

*From 29 CFR 1910.269 Tables R-6 & R-7 altitude corrected (R-5) for 1,500 m, 3,000 m, and 4,200 m.
Reprinted, by permission, from American National Standards Institute 2017a.

Figure 1.3 Job briefings should be performed before the start of each job.

Wright Tree Service, Inc.

STEP ONE-DISTRIBUTION

Distribution Job Briefing/Hazard Assessment

Always Immediately Report: Events/Incidents, ALL Vehicle Accidents, ALL Injuries

Circuits/Feeder/WO/#: _____   Date: _____ Time: _____

GPS Location ((if available) _____

General Foreman: _____   Phone #: _____

Foreman: _____   Crew Contact Ph#: _____

WTSW Job #: _____   Equipment #: _____

Job Description / Overview: _____

	INITIAL MORNING	INITIAL AFTERNOON

HAZARDS ASSOCIATED WITH THE JOB
Check ALL identified hazards

- ☐ Inspected tree root collar health?
- ☐ Inspected tree trunk health?
- ☐ Inspected tree limb health?
- ☐ Identified decay or dead wood?
- ☐ Identified weak crotches?
- ☐ Is there tension wood (stored energy)?
- ☐ Identified tree overhang?
- ☐ Can vines be trimmed safely?
- ☐ Is the tree strong, weak, fibrous, or brittle?

- ☐ Is there fungi or mushrooms present?
- ☐ Are the limbs touching the primary or secondary?
- ☐ Are the limbs in-between the wires? (Primary or Secondary?)
- ☐ Self made hazards? (housekeeping)
- ☐ Identified Bees/Hornets' Nests?
- ☐ Are there slip, trip, & fall hazards?
- ☐ Is there pedestrian traffic?
- ☐ Is there vehicular traffic?

- ☐ Is there mud, ice or snow?
- ☐ Is there uneven terrain?
- ☐ Are there warm weather conditions?
- ☐ Are there cold weather conditions?
- ☐ Are there wet working conditions?
- ☐ Is it safe to trim without causing outage?
- ☐ Poison Ivy/Oak, thorny vegetation?
- ☐ Is there faulty Utility equipment?
- ☐ Pre-existing property damage (must be reported)

Other hazards (Give Details): _____

OBSTACLES AND SPECIAL PRECAUTIONS
Check ALL that apply

- ☐ Verified correct circuit/work location?
- ☐ Approved by owner/co-owner/landlord?
- ☐ State & County permits obtained?
- ☐ De-energization requested or needed?
- ☐ Vegetation or landscaping at risk?
- ☐ Permission to drive on private property?

- ☐ Proper lane closure in place?
- ☐ Flagging procedures understood?
- ☐ Cones spaced 1 ft per 1 mph?
- ☐ Signs placed 10ft x MPH?
- ☐ Wheel chocks in place?
- ☐ Rescue equipment in place?

- ☐ Wildlife / Plant protected area?
- ☐ Adequate lighting? (dusk/dawn)
- ☐ Obstacles & Hazards marked with Tape?
- ☐ Strobe Light works (circle) yes / no
- ☐ Secured Load if necessary?
- ☐ Utility lockout tag present?

Other Special Precautions (Give Details): _____

PLANNING THE WORK PROCEDURES INVOLVED
Check ALL that apply

- ☐ Emergency Action Plan agreed on?
- ☐ Proper placement of all equipment?
- ☐ Selected proper crotch?
- ☐ Discussed proper lifting?
- ☐ Reviewed 2 way Communications?
- ☐ Reviewed hand signals?

- ☐ Reviewed Job assignments on site?
- ☐ Rigging & roping procedures discussed?
- ☐ Roping plan established?
- ☐ Reviewed 5 step felling plan?
- ☐ Climbing Route established?
- ☐ Figure 8 knot needed at end of rope?

- ☐ Set up for safe chipper operation?
- ☐ Set up for safe bucket operation?
- ☐ Stretch & Flex completed?
- ☐ Do I need a second opinion?
  GF _____ Foreman _____ Other _____

- ☐ Handsaw, Chainsaw, Hydraulic saw: When cutting: are size, weight, and length manageable? If you lose control, where will it fall? Stop & Rethink - Do I need to do it differently?
- ☐ Pruner: determine size of cut. Is it brittle? Will I be able to maintain control while using the pruner? Where will it fall if something goes wrong? Stop & Rethink - Do I need to do it differently?

Other Plans (Give Details): _____

EQUIPMENT INSPECTION AND USE (PPE, trucks, tools, etc.)
Check ALL that apply

- ☐ Crew Personnel PPE, need replacement?
- ☐ Climbing gear inspected?
- ☐ Daily boom inspection completed?
- ☐ Daily Bucket Pre-flight?
- ☐ Bucket fall protection inspected?
- ☐ Chipper inspection completed?

- ☐ Flagging equipment inspected?
- ☐ Equipment-keys out & chocks placed?
- ☐ Tools to be used inspected?
- ☐ First Aid Kit - complete yes / no _____
- ☐ Fire extinguisher - charged yes / no
- ☐ Storm Light - (circle) yes / no

- ☐ Chaps?
- ☐ D.O.T. Inspection Completed?
- ☐ Emergency Safety Bag?

Other inspections (Give Details): _____

DISTANCE: Minimum Approach: _____   Max. Named Voltage(s): _____

Wright Tree Service

Figure 1.4 Checklists can be used as part of pre-work briefings to review job details, including hazards and procedures.

Copyright © 2018 by International Society of Arboriculture. All rights reserved.

Other Applicable American National Standards

Fall Protection Systems for Construction and Demolition Operations (ANSI/ASSE A10.32-2012)

Gasoline-Powered Chainsaws (B175.1-2000)

High-Visibility Safety Apparel and Accessories (ANSI/ISEA 107-2015)

Industrial Head Protection (ANSI/ISEA Z89.1-2014)

Ladders - Portable Metal (ANSI ASC A14.2-2007)

Ladders - Portable Reinforced Plastic (ANSI ASC A14.5-2007)

Ladders - Wood (ANSI ASC A14.1-2007)

Mast-Climbing Work Platforms (ANSI/SIA A92.9-2011)

Occupational and Educational Personal Eye and Face Protection Devices (ANSI/ISEA Z87.1-2015)

Personal Fall Arrest Systems, Subsystems and Components (ANSI/ASSE Z359.1-2007)

Respiratory Protection (ANSI/ASSE Z88.2-2015)

Tree, Shrub, and Other Woody Plant Management (ANSI A300 Parts 1–9)

Vehicle-Mounted Elevating and Rotating Aerial Devices (ANSI/SAIA 92.2-2015)

Workplace Walking/Working Surfaces and Their Access; Workplace, Floor, Wall and Roof Openings; Stairs and Guardrails Systems (ANSI/ASSE A1264.1-2007)

Some specific cases of concern follow. Note that they do not cover all potential electrical safety risks an arborist might encounter, and this section is not intended to substitute for a comprehensive incident prevention program.

Tree on Line

If incident prevention fails and a tree or tree part falls on a line, it can be very dangerous. The best course of action is to immediately call the operating utility control center and, if necessary, emergency crews. Maintain minimum approach distances for everyone. Respect approach distances from anything that is or could be energized, including a tree, fence, tool, or another object. In order to avoid the risk of step potential, stay at least 35 feet (10.7 m) away from the base of the tree, tree part, or other potentially energized object on the ground. If someone is in the tree, still alive and not being shocked, they are probably best to stay put until the utility arrives to shut off power, rather than run a gauntlet of electric hazard to evacuate the tree.

On the other hand, if the person is electrified, there is not much anyone can do to mount a rescue without risking his or her own life. If there is enough energy present to threaten one person, it is a good bet there is enough to threaten others. Remember that an energized human body, tree, or climbing rope are all indirect contact threats. Potential rescuers should not touch anything that is or could be energized. Electrical contact will thwart the rescue by injuring or killing the rescuer. That does not help the original victim and compounds the tragedy by increasing the number of casualties. In nearly all cases where someone is energized, the only reasonable course of action is to wait for utility and emergency crews to arrive. What is surprising to many people is that if the worker is not immediately electrocuted by the contact, they often survive to be rescued 20 to 40 minutes later after the line has been properly de-energized and emergency medical technicians arrive. However, while they may live, they will still suffer serious burns and other injuries. So preventing an electrical contact incident is always the best approach.

Protective Grounds

High voltage lines should be considered energized unless visible, protective grounds have been installed and are under the supervision of a journeyman lineman. Protective grounds provide defense against inadvertent re-energization, which can result from back feed, contact with neighboring circuits, or lightning. Protective grounds also protect from induction risk. Even after protective grounds have been installed, a current could build due to back feed from a generator, back feed from within the electrical utility system, or induction from neighboring circuits.

During re-energization, grounds cause a short circuit, directing the current to ground. Lines should not be considered de-energized unless visible grounds are present. For intact lines, a single set of grounds is considered sufficient by most utilities. Called single point grounding, it involves setting grounds between the conductors, the system neutral (on Wye configuration), and the ground (a tower or pole ground) that establishes an "equipotential zone," which is a work area at a near identical state of electrical potential and is safe to work. If lines are downed, a set of grounds on either side of the break may be required. Known as bracket grounding, it involves installing one set of grounds on either side of a work site (OSHA n.d. [c]). Arborists should be aware that a common cause of incidents has been neglecting to ground across open points in a conductor, such as when lines are broken, even when grounds are installed on both sides of the break or open point.

Downed Lines

Downed lines are dangerous because they are accessible for direct contact to people on the ground due to step and touch potential. They can fall across and energize conductive objects, such as fences, vehicles, or metal-sided buildings. In the case of fences, the threat of indirect contact could be expanded exponentially as energized fences could connect entire neighborhoods, potentially endangering people in their yards throughout an entire area. Moreover, live wires can arc and whip around, possibly striking bystanders and inflicting physical or electrical injury.

Lines down across occupied vehicles are special cases. In these instances, people in the vehicle are usually safest to stay inside. Electricity will flow like water off the vehicle to ground and ordinarily pose little danger to vehicle occupants. Call the utility and emergency crews, and keep all others at least 35 feet (10.7 m) away. Do not let anyone touch or stand near the car or truck. Making contact with the ground and an energized object at the same time will provide a path to ground for electricity through the body due to touch potential.

There are few circumstances where occupants of an energized vehicle are safer to evacuate than to stay inside. Fire is one of those cases. If occupants have to evacuate, they must jump clear, making sure to avoid touch potential by landing without contacting the vehicle, and to evade step potential by landing with their feet together without falling down. They don't need to jump far, just far enough so they can land without touching the vehicle. They must clear minimum approach distances by hopping, keeping feet together and shuffling, or bounding, touching only one foot to the ground at any one time. This is a difficult task to complete, and the outcome of failure is often death. So affected motorists should remain in place unless there is a serious immediate hazard to avoid, such as fire.

Back Feed

Back feed is electrical flow in an unintended direction through a transformer. While transformers on distribution systems are most commonly used to step down voltage from high voltage to household voltage, the same transformers can step up from the home into the distribution system through back feed (see Chapter 5). For example, customers often use home generators during outages. Moreover, many homes are now wired for their own wind turbines or solar cells, and these generators can also back feed into the electrical grid. If a running generator or other electrical source is not appropriately isolated from the grid, electricity from the generator can feed into a distribution transformer, which will "step up" that voltage from

120 volts to distribution potential. Bystanders or workers near what they think are de-energized, downed lines could be injured, maimed, or killed as a result. Consequently, electrical lines should never be considered dead unless a trained lineman has grounded them and advises that they are safe.

CREATING A CULTURE OF SAFETY

A safety culture integrates the employer's responsibility (to provide a workplace free of unsafe conditions) and the worker's responsibility (to minimize unsafe acts) with behavior-based safety training and the practices of high-performing organizations into effective incident prevention. It adopts a belief system that the most important goal of every work day is for everyone to return home safely. It is both learned and taught, and has to involve every team member, from senior management to the newest groundworker. That means everyone has to be a leader and mentor and to be approachable with both positive and negative feedback. It is important to develop a culture where everyone accepts responsibility for one another's safety (Nutter 2012). A culture of safety is dependent on sound leadership using the techniques described in Chapter 2.

Senior management must be committed to safety and provide leadership by example. They have to encourage workers to report close calls without fear of punishment. These are opportunities to improve safety within a company, not times to point fingers and lay blame. Management should establish positive accountability by building relationships, using science to understand at-risk behavior, maintaining a safe physical environment, and creating systems that encourage safe behavior (Agnew and Daniels 2015). It is also important for them to make appropriate hiring choices so they don't subject current employees to careless or inconsiderate colleagues. That means taking the time to check references and conduct pre-employment screening of past employment history and arrest records (Gober 2012). Managers should openly share the company's safety goals and successes, explain the consequences of carelessness, and reward safe behavior. Rewards need not be elaborate. It is often sufficient to simply provide recognition for safe, hard work. Training is essential, and it must be ongoing. Utilities and contractors should collaborate with one another for training sessions (Figure 1.5) (Nutter 2012).

Formal job behavior observations should be conducted by management and workmates alike. Job behavior observations document safe and unsafe acts to help everyone understand whether or not the group is trending appropriately. They can be used in the safety matrix suggested by McClenahan (2012).

Ultimately, effectively creating a culture of safety combines the various safety philosophies. Behavior-based theory, multiple causation theory, and high reliability organizational characteristics should all be integrated into an effective safety program. Unsafe acts and conditions lead to incidents, and it makes sense to focus on work conditions and behaviors that present greater risk of injury. Moreover, preoccupation with failure, elaborating on close calls, emphasizing operations, resiliency, and giving deference to expertise should all be inherent parts of a sound safety program. New or modified safe behaviors will invariably develop. These new behaviors should be targeted for positive habit development. Once

Figure 1.5 Training sessions are part of a culture of safety.

unconscious competence evolves, these new behaviors then become part of the safety culture (Agnew and Daniels 2015).

SUMMARY

Tree work has unique potential risks due to the heavy loads, work height, power equipment, and extreme weather conditions associated with the trade. Incidents are unplanned, undesired events that could lead to injury. The three most common types of fatal incidents associated with tree work are contact with objects or equipment (struck-bys), falls, and exposure to harmful substances. Electricity is considered a harmful substance.

Employers have a moral and legal responsibility to provide a workplace free of known hazards. Developing a culture of safety requires sound, positive leadership using techniques described in Chapter 2. Safety cultures can be achieved through training, safety committees, and by applying the same communication, responsibility, authority, and accountability tools that have historically proven effective to control quality, productivity, and cost. Workers also have a responsibility for their own safety and the safety of their colleagues. They have to understand and comply with corporate work standards and rules, governmental regulations and other standards, follow their supervisor's instructions, and participate in training.

Behavior-based industrial safety principles were pioneered in the 1930s by Herbert Heinrich. While they have been modified by Petersen, McClenahan, Agnew and Daniels, and others, the underlying principles of Heinrich's work still apply today. Heinrich reached a conclusion that the more frequently a labor force works unsafely and is exposed to unsafe conditions, the greater the chance they will have an accident (Heinrich's term). Heinrich determined that the best way to prevent serious or fatal incidents is to eliminate unsafe acts and conditions. He illustrated the concept with an accident pyramid in an example comprising untold thousands of unsafe acts or conditions at the base of the pyramid, which led to hundreds of close calls, tens of minor injuries, a number of serious injuries, and ultimately a fatality at the top. Fatalities, while unacceptably high in our industry, are still relatively rare events, but when one occurs, it is common to find that the company had already experienced a number of serious, non-fatal incidents. They had been working their way up the pyramid. A series of close calls over a short period is a wake-up call that something is amiss within the company's safety program that, unless identified and corrected, may result in injuries.

Electrical injuries are serious, and high voltage contact is either fatal or life-altering, which is a major reason why incident prevention is so critical in utility arboriculture. ANSI Z133 is the operative safety standard in the United States that has the force of law insofar as it is a document often referenced by OSHA compliance officers in issuing a citation for safety violation for tree care operations. Careful adherence to this document is an essential step in fostering a culture of safety within a company.

A safety culture integrates the employer's responsibility to reduce unsafe conditions and the worker's responsibility to reduce unsafe acts through behavior-based safety training and the practices of high-performing organizations into effective incident prevention. It fosters a belief system that the most important goal of every work day is for everyone to return home safely. It is both learned and taught, and has to involve every team member, from senior management to the newest groundworker. That means everyone has to be a leader and mentor and to be approachable with both positive and negative feedback. In a culture of safety, everyone accepts responsibility, not only for themselves, but also for one another. Positive accountability is needed to achieve a high-performing safety culture. It involves relationship development, using science to understand at-risk behavior, maintaining a safe physical environment, and creating systems that encourage safe behavior.

CHAPTER 1 WORKBOOK

Fill in the Blank

1. In the United States, enforcement of safety work rules resides with the _____ ___ _____ _____ (____).

2. A worker neglecting to wear cut resistant chaps while operating a chainsaw is committing an unsafe _____, while an employer failing to provide chaps is creating an unsafe _____.

3. The ___-_____ rule suggests that workers should take time before starting a job to check for potentially unsafe conditions and confer about them.

4. According to ANSI Z133, the minimum approach distance for persons other than qualified line-clearance arborists for 7.2 kV is ___ feet / ____ m.

5. Surface temperatures at the point of contact with high voltage can reach _____°C / _____°F.

6. _____ _____ occurs when someone touches an energized fixture. _____ _____ can occur when a person touches a conductive object that is in contact with an energized fixture.

7. The difference in voltage between two objects someone simultaneously contacts is termed _____ _____.

8. _____ _____ is the difference in voltage on the ground between two of a person's body parts near where electricity goes into the ground.

9. An energized human body, tree, or climbing rope are all examples of _____ _____ threats.

Multiple Choice

1. The safety acronym S.T.A.R is valuable for preventing errors, and stands for
   a. Stop, Think, Act, Review
   b. Stop, Think, Adjust, Reason
   c. Safety, Training, Action, Review
   d. Stop, Train, Act, Reflect

2. A high-reliability organization would *not* always accept the
   a. simplest explanation for close calls or minor incidents
   b. need to adhere to electrical minimum approach distances
   c. employee's first request for personal protective equipment
   d. necessity to perform safety meetings and work briefings

3. There are no published safety data related specifically to utility arboriculture.
   a. True
   b. False

4. Which of the following is *not* a recommend way for an employer to create a safe workplace?
   a. provide a workplace that is free of known dangers
   b. provide personal protective equipment free of charge
   c. provide employees with minimal control over their work environment
   d. inform employees of their rights and duties under the company's safety and health program

5. Unsafe acts include which of the following?
   a. willful rule violation
   b. maliciousness
   c. human error
   d. all of the above

6. The term *accident* is in disfavor because it implies that outcomes are due to
   a. equipment failure
   b. ignorance
   c. bad luck
   d. lack of training

CHALLENGE QUESTIONS

1. Using the tools to improve human performance identified by the Tennessee Valley Authority, describe ways your organization could improve safety on the job site.

2. Why is an over-reliance on discipline counterproductive to a safety program?

3. Why should arborists be aware of back feed, and under what circumstances does it occur?

4. Describe the appropriate course of action for "tree on line" scenarios when (a) no one is in the tree, (b) when someone is in the tree and the system is not electrified, and (c) when someone is in the tree and the system is electrified.

5. Explain why safety committees are important for improving and maintaining job site safety, and identify steps to take when organizing a safety committee.

2
Program and Personnel Management

OBJECTIVES

- Develop a strategic plan for a vegetation management program.
- Design work schedule plans that best achieve program goals and objectives.
- Utilize project management techniques to execute project plans.
- Prepare a utility vegetation management budget.
- Execute contracts for vegetation management services based on needed services and budget.
- Implement a personnel management strategy that encourages high-performing staff.

KEY TERMS

- budget
- capital budget
- circuit work
- constraint triangle
- contract
- cost center
- critical path
- decision package
- dependencies
- evergreen contract
- geographic information system (GIS)
- grid work
- hotspotting
- key performance indicator (KPI)
- labor union
- lump sum contract
- mid-cycle pruning
- operating and maintenance budget
- performance appraisal
- performance-based contract
- performance budgeting
- predictive maintenance
- profit center
- program budgeting
- preventive maintenance
- reactive maintenance
- specifications
- SMART goals
- SWOT analysis
- time and material contract
- triple constraint
- unit price contract
- vehicle monitoring system (VMS)
- work breakdown structure

INTRODUCTION

Utility vegetation programs are often large and complex organizations. Managers with both utilities and contractors require a thorough knowledge of communication, planning, writing and maintaining specifications, project management, setting and keeping on budgets, personnel management, contracting and procurement, fleet management, compliance, and other topics. This chapter will touch on these and other subjects. Readers should understand that each of these topics discussed in detail could fill a book by itself, and those interested should pursue specific subjects in greater depth.

PROGRAM PLANNING

Miller and colleagues (2015) observe that planning takes place in both short- and long-term time horizons. Long-term planning is strategic. It involves establishing overall goals and objectives, prioritizing desired outcomes, and establishing action plans to achieve them. Short-term planning is tactical, and focuses on day-to-day management activities, such as scheduling tasks, equipment maintenance, supervision, reporting, record keeping, and action plans to deal with potential crisis situations. The adaptive management planning process in integrated vegetation management presented in Chapter 4 is tactical in scope, but it can be applied strategically as well. The benefit of the process is that it is documented, methodical, and structured to achieve successful results. It enables managers to anticipate and correct problems before they arise, and should be based on up-to-date information (Miller 2014).

SWOT Analysis

A Strengths, Weaknesses, Opportunities, and Threats (**SWOT**) analysis is an effective strategic planning tool often presented in a matrix (Figure 2.1) (Berry and Wilson 2005). The analysis can be completed by answering questions in each matrix category. Performed properly, SWOT analysis helps organizations focus their strengths to develop effective objectives that lead to success (Berry and Wilson 2005).

Specifications

Specifications are documents that describe, instruct, and direct work results. They should be written carefully since they become part of legally binding contracts (Matheny and Clark 2008). Moreover, well-written specifications minimize misunderstandings that could otherwise result in conflict and undesirable results.

Strengths	Weaknesses
What do we do well?	What do we need to improve?
What unique resources can we leverage?	Where do we lack resources?
What do others recognize as our strengths?	What are others likely to view as our weaknesses?
Opportunities	**Threats**
What opportunities are presented to us?	What potentially harmful threats loom?
What trends are available to us?	What are our competitors doing?
How can we convert our strengths to opportunities?	To what threats do our weaknesses expose us?

Figure 2.1 A SWOT analysis matrix. *(Berry and Wilson 2005; MindTools 2011)*

Specifications may cover elements detailing safety, scheduling and reporting, work practices, performance standards, customer relations, environmental protection, chemicals, equipment type, workmanship, and other topics. Specifications based on applicable standards (such as the ANSI A300 *American National Standard for Tree Care Operations* series, particularly Parts 1 and 7 [ANSI 2017b, ANSI 2012] and ANSI Z133 *American National Standard for Arboricultural Operations – Safety Requirements* [ANSI 2017a]) and best management practices (such as ISA's Best Management Practices Series, including *Utility Pruning of Trees* [Kempter 2004] and *Integrated Vegetation Management* [Miller 2014]) have credibility and authority.

Work Scheduling

Proper work scheduling accommodates program goals and objectives (Miller et al. 2015). Resulting schedules should be aligned with reporting capabilities. There are three different approaches to work planning: cycle-based, results-based, and crisis management.

UVM efforts can be preventive or reactive. As the names imply, **preventive maintenance** is performed before an occurrence (such as a tree-caused service interruption), and **reactive maintenance** happens in response to an occurrence (Odom 2010). Reactive maintenance is inherently inefficient. However, preventive maintenance can also be inefficient, especially when resources are expended to prune or remove trees that are unlikely to cause problems. To ensure the best use of resources, it is good practice to identify vegetation that is most likely to create difficulty prior to assigning work. Since performing inspections is less costly than actually performing work, a combination of periodic inspections and maintenance pruning as needed is often the most effective strategy.

Maintenance Intervals (Cycles)

The maintenance interval (sometimes called the maintenance cycle) is the planned period between line-clearance operations. However, tree populations are subject to dynamic environmental conditions, such as storms, precipitation patterns, and pest infestations (Utility Arborist Association 2014). For this reason, the concept of **predictive maintenance** (also known as "on-condition maintenance"), which directs resources based on actual field conditions, has gained favor (Hashemian and Bean, 2011). Regardless of whether maintenance is performed on a cyclical or predictive basis, how maintenance resources are allocated for any specific site is determined by taking into account factors such as:

- Tree species mix, age, condition, and expected growth rates
- Community norms
- Typical weather patterns
- Regulatory requirements
- Priority of the facility
- Available funds

Past maintenance practices, including the specification and quality of work, can influence tree growth rates, risks posed, and the cost and approach to future maintenance efforts (Grayson n.d.). For example, poorly placed cuts, excessively large cuts, or topping may result in rapid growth or weaker tree structure and increased rates of tree failure. Likewise, excessively long maintenance intervals result in the need for heavier pruning, which can negatively affect tree health, increase tree risk, and induce excessive sprout growth (see Chapter 3).

Governments sometimes mandate clearance distances, maintenance intervals, inspection intervals, or require available resources to be focused first on "worst-performing circuits" (CNUC 2010). These rules can affect how utility inspection and pruning are performed. For example, due to the high risk of fire in the state of California, regulated utilities face stiff fines for non-compliance with clearance requirements. As a result, most overhead lines in that state are patrolled at least annually, and fast-growing trees are pruned more frequently than those with slower growth rates (S. Tankersley, personal communication).

Determining Optimum Maintenance Intervals

In general, inspection and maintenance should be performed more frequently in areas with long growing seasons or with a high percentage of fast-growing tree species. Likewise, in areas with short growing seasons or populations of many slow-growing tree species, intervals tend to be longer (Sankowich 2013). However, while a baseline maintenance interval may be stated for an entire system, lengths may vary locally depending on factors such as those listed above (Utility Arborist Association 2014). Ideally, utility pruning should be performed at intervals that minimize the need for reactive responses, but not so frequently that costs exceed benefits (Campbell 2012).

Relatively short inspection and maintenance intervals are more effective in achieving reliability and safety objectives, including improved storm performance. In addition, shorter intervals reduce the biological impact on trees and the aesthetic impact on neighborhoods, which can reduce the number of customer complaints (Odom 2012). Therefore, determining optimum cycle length for any particular location will depend on a comprehensive analysis of the costs and benefits of the spending, including not only the cost of operations and repairs, but also the value of reliability to utility customers and the benefits of better customer relations (Sankowich 2013; Utility Arborist Association 2014).

The effectiveness of cycle-based maintenance can be enhanced by basing specifications on expected tree growth rates rather than a fixed distance for all trees (Cieslewicz and Novembri 2004). Fast-growing trees that would otherwise shorten maintenance intervals are sometimes controlled with **mid-cycle pruning**. These programs target trees that are growing significantly faster than most of the other trees on a scheduled pruning cycle, enabling a longer interval for slower-growing trees (Sankowich 2013).

Maintenance intervals may vary between urban areas, where comparatively large clearances may not be achievable for aesthetic reasons, and rural areas, where landowners may have fewer concerns and greater clearances are acceptable (Figure 2.2). Likewise, areas more prone to severe weather patterns often require greater clearances and/or shorter maintenance intervals to reduce the impact of storms (CNUC 2010a).

CNUC (Cieslewicz and Porter 2010) found a five-year average distribution cycle length among North American utilities in 2010. Fifty-two percent of utilities that reported to be working on a cycle had changed their cycle length since 2006, with the average change increasing from 4.5 to 6.6 years. However, only 45 percent of respondents were on a cycle. The reasons companies cited for not maintaining cycles varied from insufficient money to transitioning to performance-based contracts.

Figure 2.2 Greater clearances may be more consistently obtained in some rural areas.

Implementing Predictive Maintenance

Predictive maintenance focuses resources on areas where maintenance is most needed, based on facility priority and data about actual field conditions (Hashemian and Bean 2011). Field data can be gathered with trained ground patrols, using tree risk assessment methodology, and supplemented with aerial patrols, such as LiDAR and aerial photography. Pruning operations that are part of a predictive maintenance regimen should be performed preventively—i.e., before problems occur—and implemented systematically, to maximize productivity.

These enhancements require greater professional expertise for inspectors, work planners, and crews. However, the value added by improving overall program effectiveness, including better customer relations, far exceeds the costs of workforce professional development (S. Tankersley, personal communication).

Just-In-Time Management

Just-in-time management is a predictive strategy designed to target trees as close as possible before they interfere with facilities. It is a modification of **hotspotting** (addressing only those trees that are going to cause immediate difficulties) insofar as it covers an entire system in a specified time period, but only addresses trees that are anticipated to contact electric lines over that cycle. The idea is to increase efficiency by only working trees that are most likely to cause problems in the near term. The most effective just-in-time programs also have proactive sapling and tree removal components to reduce long-term work volume.

A difficulty with just-in-time management is that it provides little or no margin for error. If a program does not have accurate work volume information, suffers a budget cut—for even a year—or lacks a proactive removal program, unmanageable conditions may develop, and the program could quickly devolve into crisis management. Another associated difficulty is that it requires considerable expense to patrol, survey, record, catalog, and map trees that will need maintenance in a particular year (Cieslewicz and Novembri 2004).

Reactive (Crisis) Management

Reactive management develops when the majority of work is done in reaction to undesirable conditions. Undesirable conditions lead to unacceptable service reliability and electric safety risks. Examples include an unmanageable number of trees interfering with electric lines, a backlog of defective trees that are an immediate threat to facilities, trees that could start fires, and other issues. These conditions often generate customer service requests that demand attention, which further increases unscheduled workloads.

While at first reacting to unscheduled work may appear responsive, it is inefficient. Reactive work can cost five times as much as doing the same job as part of routine maintenance (J. Downie, personal communication). A major reason for the increased cost is that reactive locations are rarely near work sites where crews are deployed at a given moment. Pulling off tree crews to respond to a crisis in such cases often means they have to drive considerable distances to their new assignments. The additional hours behind the wheel detract from time available for tree work, which reduces productivity.

Emphasis on reactive work makes it nearly impossible to address long-term goals (Matheny and Clark 2008) and ultimately creates a spiral of decline. The more off-cycle work a utility performs, the less scheduled work gets done. Less scheduled work allows other potential threats to develop, which demand even more reactive work, so conditions deteriorate further and so on, and the system ultimately devolves into crisis management.

Crisis management is an indicator of chronic underfunding (Miller et al. 2015). Recovery demands high levels of resources to be applied to scheduled work and discipline to minimize off-cycle assignments to only those necessary to effectively manage risks. Utilities that rely on reactive maintenance experience greater long-term cost, increased tree-related outages, and heightened safety risks, which ultimately lead to public outcry and increased media scrutiny with accompanying reputational censure (Oberbeck and Smart 2004).

Circuit vs. Grid Work

Vegetation management operations are often conducted on a circuit or grid basis. Both circuit and grid philosophies can be effective. The circuit approach is common on transmission as well as rural and suburban distribution projects. Systematically completing entire circuits can help minimize tree-related outages because all of the line is clear by the time the project is completed. **Circuit work** also allows for a direct correlation to be made between vegetation management efforts and electric system reliability.

Grid work consists of working all of the line within a geographic boundary. It can be more practical on urban distribution facilities where circuits may be periodically reconfigured to accommodate load adjustments. If the area is managed on a circuit basis, reconfiguration can put some sections of line out of sync with others, or even leave them neglected if the changes are inappropriately tracked. A grid system provides a practical means of making sure such line segments are not missed. A disadvantage of grid work is that circuits often cross grid boundaries. If adjacent grids are not worked in a timely manner, some portions could be left with conditions that negatively affect reliability.

Regardless of the scheduling strategy adopted, thorough, systematic work is essential for success. If a system is on plan, either a grid or circuit approach can be effective. If not, neither will prevent problems from developing.

PROJECT MANAGEMENT

The purpose of well planned and executed project management is to provide a consistent, systematic approach to meeting program objectives. Vegetation managers should be aware that there are many helpful project management software packages that are available for their use.

Triple Constraint

Baca (2007) observes that projects are subjected to a **triple constraint**, consisting of time, cost, and scope. These elements can be depicted in a **constraint triangle** (Figure 2.3). Time, cost, and scope

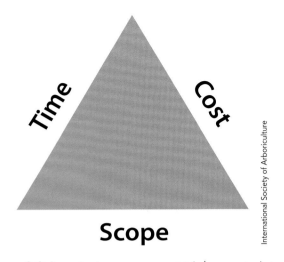

Figure 2.3 A project management triple constraint triangle. Time, cost, and scope compete with one another for limited resources and have to be balanced for successful project completion.

are interrelated insofar as they compete with one another for limited resources and have to be balanced for successful project completion. For example, if a project deadline is tight, a manager will need sufficient resources to deliver it on time. However, resource limitations might make an ambitious deadline unachievable. If both time and money are constrained, the scope of the project has to be modified if the project is to be completed as scheduled and on budget.

To illustrate how triple constraints might work in a utility vegetation management context, consider a transmission project that is constrained by time because it must be completed in the coming calendar year for compliance with FAC-003. The planned scope is to clear the right-of-way to a full wire-border zone prescription (see Chapter 4). That's fine if there is sufficient money for the human and equipment resources needed to finish the job to full specifications by year-end. If not, additional money needs to be applied to the project or funding also becomes limiting. If money presents a second constraint, the scope will need to be altered to meet the time objective. In this case, the scope might be modified to hotspotting, or some other approach. The point is that if one or more constraints are limited, an adjustment has to be made in one or two of the others or the project will fail.

Work Breakdown Structure

A **work breakdown structure** is a prioritized outline of job components that must be completed for a successful project outcome (Baca 2007). The work breakdown structure serves as a plan framework and can be lined out using project management software. The longest period of time a series of tasks requires from project beginning to end is the **critical path** (Kendrick 2004).

Work plans document task beginning and end, assign personnel, designate responsibility, relate specifications, and itemize costs (Matheny and Clark 2008). The highest level is the project overview, and intermediate levels are waypoints toward completion. The final level is the work package that brings the project to a close. Work packages are often assigned to a single individual or team of individuals, like tree crews (Baca 2007).

Dependencies

Dependencies are the relationships that dictate when tasks in the work breakdown structure begin and end. Baca (2007) identifies three types: mandatory, discretionary, and external.

- A mandatory dependency is one in which one task cannot begin until another activity has ended. For example, some utilities may not allow a contractor to begin work on a project without written authorization. In this case, field work is mandatorily dependent on written authorization from the utility.
- Discretionary dependencies are developed by managers to meet specific objectives. For example, a vegetation management team might decide to complete distribution three-phase lines before they begin single-phase taps. In such a case, a discretionary dependency exists between single- and three-phase work.
- An external dependency is a relationship between a task under the direction of a project management team and an action outside of their control. For example, vegetation management work on federal lands in the United States requires written notice from governmental land managers before it can proceed. Since utilities cannot initiate work until permitted, federal authorization to begin work is an external dependency. This is also an example of a mandatory dependency.

Dependencies have relationships to one another relative to when they begin and end (Baca 2007). For example, a finish-to-start dependency describes a relationship where a preceding task must be fully completed before its successor can begin. The discretionary dependency that requires three-phase distribution lines to be finished before crews begin work on single-phase lines is a finish-to-start relationship. In a start-to-start dependency, the successor can't start until the predecessor has started. A start-to-start case might involve pre-inspectors, where line-clearance work can't start until pre-inspection starts. In a finish-to-finish dependency, a successor cannot finish until a predecessor has finished. For instance, some utilities distribute surveys to customers on whose property line-clearance work was conducted. Survey cards can't be returned until after work has been completed. The final relationship is a start-to-finish dependency, where the successor cannot finish until a predecessor starts. Start-to-finish dependencies are rare (Baca 2007).

Gantt Charts

Gantt charts, which were developed by Henry Gantt in the early 1900s, are horizontal bar graphs plotted along a timeline. Gantt charts typically list activities or tasks associated with a project along the Y-axis and time along the X-axis (Figure 2.4). They are scheduling tools that provide a visual means of tracking a project, representing the desired starting date for every task involved in a project and the estimated length of time each will take (Matheny and Clark 2008).

Project management software will not only plot Gantt charts, but also work breakdown structures, critical paths, dependencies, scopes (e.g., line miles or acres), start and stop dates, budgeted monetary resources, and other relevant factors. The software may be used for individual projects, such as a line or grid, work over an entire year (or other specified length of time), or both. The visual layout allows managers to view dependencies

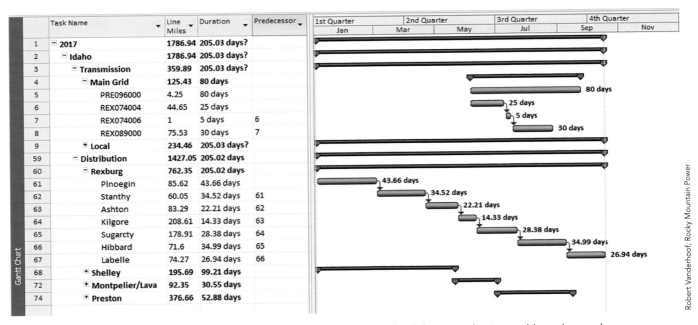

Figure 2.4 Gantt charts, which are used in project management scheduling, are horizontal bar charts along a timeline.

and anticipate constraints so they can be proactively addressed (Kendrick 2004).

Work Planners

In a 2006 industry survey, Cieslewicz and Novembri found that 55 percent of utility vegetation management programs contracted pre-inspectors; by 2012 the figure had increased to 59 percent (CNUC 2015). Ideally, work planners should be familiar with all facets of vegetation management and are not only involved in organizing work for tree crews, but also in customer relations, including working through property owner resistance to tree work. In most cases, they are perceived as the face of the utility, interacting with more customers than any other utility representatives. In organizing work, they can note such information as the name of the project (e.g., transmission line, distribution circuit, or grid identifiers), property owner identification (if available), crew type most suitable for the location, a description of the work, and any other noteworthy details. Work parameters could include the location, species, condition, and growth rate of vegetation that needs to be cleared from the lines. Work planners might detail which trees should be pruned for clearance and identify and mark trees for removal. Increasingly, contractor and utility notification documents are created in the field with hardened laptop computers, coupled with GPS capability (Ross 2011). Tablet computers are also gaining popularity due to their affordability, portability, and functionality.

Inspectors may be contracted from third parties, be employed by the line-clearance contractor, or work directly for the utility. In a survey of vendors, Cieslewicz and Novembri (2004) found mixed perspectives on whether or not pre-inspection contracts ought to be issued to contractors other than those performing tree work. In all likelihood, the work is divided between contractors that conduct tree work and those that do not. Favorable comments in the survey for independent work planning included better communication with stakeholders, improved crew productivity (because crew leaders were relieved of the primary customer contact responsibility), higher removal rates, greater clearance, and reduced costs. The dominant opposing view was that assigning tree contractors with pre-notification responsibility left them with control over the entire operation, which was seen as disadvantageous (Cieslewicz and Novembri 2004).

Work Management Systems

Computerized work management systems integrate multiple layers of data into a **geographic information system (GIS)** platform. Geographic information systems are databases embedded in maps. Clicking on a point on a map enables access to records associated with that site, such as:

- Previous complaints or areas of concern
- A site of ongoing litigation
- Cultural sites
- "Cycle buster" trees
- Environmentally sensitive areas
- Historical lands
- Lines by type
- History: date last worked, production (e.g., man-hours a mile or kilometer, man-hours or cost of a tree, etc.)
- Governmental ownerships (e.g., federal, tribal, provincial, state, county, municipal, or other)—imbed special use permits
- Organic farms
- Chemical restrictions
- Outage locations
- Property ownerships (e.g., plat information, contact information), rights-of-way, customer history, complaints
- Riparian areas
- Roads
- Sensitive customers
- Substations
- Threatened and endangered species nesting closure and other work requirements
- Topography
- A complete tree inventory

GIS work flow systems can help managers coordinate events using data on conditions, workload inventories, ground and aerial patrol results, outages, satellite imagery, work history, property ownerships, refusal and concerned customer history, topography, environmental assessment reports, training records, surveys, testing results, and other pertinent information (Meehan 2007). They can aid in report development, preventative maintenance scheduling, and cost management, as well as tracking work history and service requests. They can also facilitate unit or cycle cost determination (Figure 2.5) (Meehan 2007).

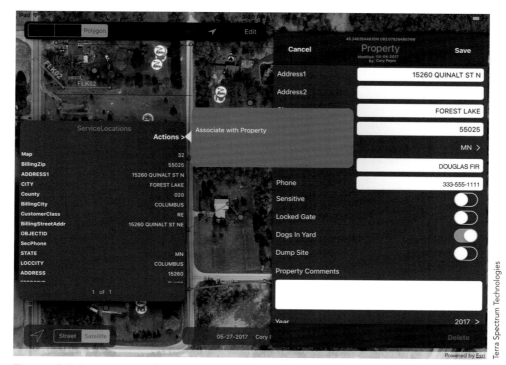

Figure 2.5 Property attributes shown on a geographic information system platform.

The visual properties of GIS capability can be valuable decision-making tools. For example, work history at a location or details on sensitive environmental or cultural sites at or near planned work can be combined with scheduling capabilities to alert managers of anticipated problems and enable them to adjust plans accordingly (Meehan 2007). GIS can also note access locations and effectively deploy crews. Vegetation types and location can be mapped, and highlighted if inside of action thresholds, so crews can be efficiently routed to where they are most needed (Meehan 2007). LiDAR data can also be applied to this purpose. Among other applications, GIS can be used in community presentations to show the consequences of failing to clear trees from power lines by establishing a relationship between outages and areas where property owners have been resistant to tree work.

In the field, GIS combined with GPS capability can increase efficiency by eliminating paper maps and leveraging the technology's data processing functions. Use of GIS can also increase efficiencies by diminishing the need to manually transfer data to reporting programs (Ross 2011).

BUDGETING

Utility vegetation management programs and contractors are often responsible for many millions of dollars annually (CNUC 2010a), yet arboricultural training and experience do not necessarily provide sufficient financial background to manage such large sums of money. Gebelein et al. (2004) recommend that managers consult with accomplished financial advisors, which is good guidance for utility arborists who are responsible for managing large budgets.

Crompton (1999) considers a **budget** to be a management plan for meeting an organization's goals and objectives. To that end, he explains that budgets are a means to justify the allocation of resources, set priorities, and make decisions regarding what can and cannot be accomplished. This process is a tool and needs to involve a level of flexibility. Shim and Siegel (1994) recommend that managers monitor their budgets, identify focus areas, and report on a scheduled basis. While the discussion in this section is not intended to replace professional assistance or provide comprehensive accounting training, it does offer a foundation that can assist utility vegetation managers in understanding how to administer their budgets.

Cost Types

Vegetation management departments for most utilities are cost centers. **Cost centers** are limited to expenditures and do not generate income. Examples of other cost centers include accounting, legal, administrative, and security departments. **Profit centers** are sources of revenue. A profit center can be a product or service, a region, or a distribution channel (Tracy 2008). Contractors have both cost and profit centers.

Variable costs change according to the level of activity during a set period of time, while fixed costs do not (Tracy 2008). For example, contractors may not know with certainty what volume of contracts they will be awarded in a given year. If they secure a substantial amount of work, they may have to add vehicles, equipment, management, and other resources—increasing costs over what they would otherwise have been. So, contractor operational costs can be variable because their workload is subject to change from one year to the next depending on the amount of sales, fuel cost, worker wages, and other factors (Tracy 2008).

Fixed expenses are costs that are "stuck" in the short term (Tracy 2008). Fixed operating expenses do not decline when sales fall. They can include office space, vehicle depreciation, existing workforce, and other line items. The only way to reduce fixed costs is to downsize the business (by laying off workers or selling office space, for example).

Fixed costs can become variable and variable costs can become fixed, depending on circumstances.

Direct costs are those that can be clearly attributed to a single organizational unit, service, or product. For example, labor for tree crew members assigned to a specific project is a direct cost. Indirect costs, on the other hand, cannot be readily associated with a specific project demand. Travel by an executive or program manager to a variety of sites for performance reviews may be an example of an indirect cost.

Preparing Budgets

Cieslewicz and Novembri (2004) observe that utility vegetation management funding levels are generally based on past budgets combined with estimates of future needs. They noted that this paradigm has been largely ineffective, as one of the most common complaints received in industry surveys is a lack of consistent and adequate funding. Properly planned and presented budgets might help remedy that pervasive problem.

Gebelein et al. (2004) recommend the following when preparing a budget:

- Review the strategic plan, set goals and objectives before beginning the budget process, and determine the resources needed to achieve program objectives.
- Carefully assess existing resources and evaluate them to determine whether or not they are appropriate for what is needed in the upcoming budgetary period.
- Review budgets and actual results both for the current year-to-date and the prior year. Look for significant variances and determine whether they should be taken into account in the new budget.
- Meet with the controller or a finance contactor to understand how your work unit's budget fits into the organization's overall processes.
  - Make note of deadlines and meet them.
  - Inquire about any general guidelines or assumptions (e.g., salary increases, expected changes in raw material prices).
  - Ask whether there are any special reports or historical financial information that could aid in the budgeting process.
- Develop contingency plans for financial constraints.
- Assign responsibility for the preparation of team or area budgets to the people accountable for them. Deliver clear instructions to those preparing portions of the budget.
- Compile all team or group budgets and reconcile differences. Make sure there are no errors of omission, particularly often overlooked items such as office equipment or supplies.

Nonfinancial managers may find it useful to identify administrative expenses by function, activity, product line, service, segment, or other responsibility center (Shim and Siegel 1994). In addition, Downie (2017) advises vegetation managers to be able to demonstrate what will be purchased for the expense (e.g., reductions in tree-related outages, wildfire risk mitigation, fewer customer complaints, and improved regulatory compliance performance) for both the long and short term.

Capital and Operating Budgets

Utilities and contractors have two predominant budgetary classifications—capital and operational. Companies often have separate budgetary and approval processes for each.

Capital Budgets

Capital budgets are monetary outlays for relatively large sums of money invested over multiple years (Tracy 2008). Capital expenditures are used to purchase fixed assets, which are durable items like buildings, machinery, equipment, vehicles, furniture, and computers. Line construction is an example of a utility capital expense. Capital budgets are usually established separately from operating budgets. Capital budgets enable organizations to analyze the capacity, condition, and efficiency of their durable assets, which helps them determine whether or not their plant and equipment need to be updated. Regulated utilities' allowable returns are based on capital assets.

Depreciation is the process of accounting for economic losses caused when capital assets wear out over time (Tracy 2008). Depreciation means that a business unit doesn't absorb the full financial burden of a purchase in the year it is acquired. Straight-line depreciation evenly spreads out costs over an item's expected life span. It involves simple division and has the advantage of stabilizing the expenses from year to year. Accelerated depreciation, on the other hand, assumes investments lose value more quickly when they are new than after they age. So, accelerated depreciation books greater amounts in the years immediately after an item is purchased than when it is older. The advantage of using accelerated depreciation is that it provides more conservative profit performance estimates than straight-line accounting, and it reduces income tax liabilities in the early life of an asset (Tracy 2008).

Operating and Maintenance Budgets

Operating and maintenance budgets finance day-to-day activities including salaries, rent, utilities, and supplies (Matheny and Clark 2008). Operating expenses can be either fixed or variable. Matheny and Clark describe three requirements for municipal operating budgets. These requirements are also appropriate for utility vegetation management programs and include:

- Addressing financial needs, resources, and plans
- Describing components of program costs, equipment, facilities, and personnel requirements
- Identifying the costs of achieving program goals and sources of revenue

Crompton (1999) identifies five operational budget formats:

Line-item budgeting. Line-item budgeting allocates funds by itemizing categories and accounts. The advantage of using line-item budgeting is that budget adjustments can be made readily. The disadvantage is that by distributing costs across different accounts, it can be difficult to evaluate how much is spent on separate programs.

Performance budgeting. Performance budgeting stresses the goals and objectives of the budgeted program. It often involves mission statements and performance measures.

Program budgeting. In program budgeting, line items are organized into programs rather than accounts. Examples of programs might be labor, chemicals, or individual projects. The arrangement focuses on outcomes, rather than inputs. Its disadvantage is that it does not concentrate on the purpose or benefits of each program.

Zero-based budgeting. Zero-based budgeting is a variation of program budgeting that is more results than financially driven. It requires creating a budget from scratch rather than using an existing budget as a foundation. Managers must evaluate the performance of their existing programs, then create and prioritize **decision packages**. Decision packages involve project expenditures, outputs, and alternatives, which receive funding in order of priority. Zero-based budgeting can be done in a strategic fashion to consider multiple years in the future. One advantage of this method is that it requires starting with a "clean-slate" each year to allow for adjustment to dynamic conditions. The disadvantage of using zero-based budgeting is the large amount of time and bureaucracy required during every budget cycle.

Entrepreneurial budgeting. Entrepreneurial budgeting is also called expenditure-control, target-based, or envelope budgeting. It is the most common budgeting type employed by vegetation management programs. In entrepreneurial budgeting, executives set a budget that allows managers to plan work in the context of company goals and policies. It provides the most autonomy and the greatest responsibility for managers.

Forecasting Revenue

For contractors and other companies with profit centers, managers need to have some idea of what their revenues will be, and that requires forecasting. Berry and Wilson (2005) advise managers to use historical data whenever possible. However, markets are dynamic, and past data doesn't necessarily predict future revenue. Adjustments may have to be made with educated guesses, which are dependent on experience and knowledge of business conditions.

Marketing for a higher price is a better sales strategy than low bidding (Berry and Wilson 2005). That can be difficult for contractors, as vegetation management service pricing is competitive. However, organizations that strategically attempt to gain market share at the expense of maintaining a sustainable profit margin will fail in the long term.

Simple Cash Accounting vs. Accrual Accounting

Simple cash accounting and accrual accounting are distinct expenditure tracking philosophies. In simple cash accounting, financial transactions are booked when they are paid, while accrual accounting records costs when they are incurred (Tracy 2008). Cash accounting is more straightforward but leaves a delay from the time transactions occur to when they are recorded. The resulting lag may tempt departmental managers to hold payments to create the false impression that their budgets are tracking better than they actually are. To remove this temptation, many companies have adopted accrual accounting. While accrual accounting is more widely accepted, accruing expenses can be complicated, as it often involves projections based on estimates and assumptions, which can create problems if they prove to be inaccurate (Tracy 2008). Utility vegetation managers should be aware of the type of accounting used in their organization and understand the potential ramifications for their budgets and programs.

Budgetary Approval

Approval is essential for even the most carefully prepared budget. Gebelein et al. (2004) offer the following suggestions to increase the likelihood of a budget being approved:

- Determine the accepted protocol for budgeting in your organization (both written and unwritten) and follow it.
- Ask others in the company about the lessons they have learned from previous budgeting experience.
- Keep your manager and financial advisor informed. Start discussions early and make sure they understand your decision making every step of the way.
- Consider how you will defend your budget. Describe the consequences if funds are cut or the ramifications if they are increased.
- Use the best, most updated data available.
- Consider how your budget will be viewed by others. For example, while it is common practice to inflate budget requests, unrealistic or indefensible submittals damage credibility, which can compromise future consideration.
- Understand how a budget fits into the larger picture for the organization. For example, hefty budget requests may be rejected if higher priority demands are drawing large amounts of resources.
- Coordinate with departments that are potential internal competitors for resources. Initiate conversations on how to share funds or work more efficiently. This will demonstrate a willingness to be a cooperative team member, which will often pay off in the future.

The ultimate measure of budgetary success for an organization should be how close they come to their budget and how well they deliver on expected operational results. Managers must be able to explain any variance. There can come a point when discrepancies are so great that a department loses credibility, which is harmful in the long term. On the other hand, an organization that dependably lands its budget will enjoy management confidence and be more likely to receive approval for reasonable requests in the future.

Budgetary Reports

Managers should periodically prepare reports to compare actual versus budgeted financial activity (Shim and Siegel 1994). Monthly, quarterly, semi-annual, and annual timeframes are common, depending on organizational objectives. Shim and Siegel (1994) recommend starting reports with a summary and following with detailed information that compares actual to budgeted figures, including explanations for variance and of trends over time. They add that managers should include schedules and tables, highlighting problem areas, possibly incorporating statistics and graphics to make important points.

Contractors should also have profit and loss reports. Profit and loss reports include information on sales revenue and expenses. Operating expenses in a profit and loss report are often itemized on the basis of purchases like salaries, rent, depreciation, taxes, insurance, and office supplies (Tracy 2008).

CONTRACTING

A **contract** is a legally enforceable agreement between two or more parties. Every contract has five basic elements: offer, acceptance, consideration, legal and possible objective, and competent parties (United States General Services Administration 2005). The *offer* is a promise of performance to approved vendors that should be made in precise terms. It can be rejected or set to expire after a period of time. *Acceptance* occurs when the contracting body agrees to purchase the product or service at the bidder's price. *Consideration* involves the price bargained and offered. The contracting body's consideration is the product or service provided by the vendor. The contractor's consideration is their monetary compensation for the product or service. The *legal and possible objective* is the purpose of the contract. If the objective is illegal or impossible, the contract is void. The *competent parties* stipulation is a requirement for both parties to have the legal capacity to enter into a contract (United States General Services Administration 2005).

In surveys of North American utilities in 2002 and 2006, CNUC (2010a) found that 94 percent of respondents contract vegetation management services. Utilities that secured contracts engaged vegetation management services like pruning and removing trees, mechanically clearing rights-of-way, and applying herbicide and tree growth regulators. CNUC (2010a) noted that the percentage of utilities contracting pre-inspection services increased from roughly 35 percent in 2002 to about 55 percent in 2006 and nearly 60 percent in 2014 (CNUC 2015). Moreover, 25 percent contracted post auditors. Others contracted managers, supervisors and clerical support, or services like inventories and research and development.

Around 60 percent of responding utilities bid out vegetation management contracts, 10 percent negotiated prices, and 5 percent assigned a contract to a single vendor. Nearly 60 percent of vegetation management contracts were for a set period of time, roughly 15 percent were on a work or project basis, and 13 percent used a combination of work and time contracting. Contract length ranged from 90 days to an indeterminate time period (i.e., an **evergreen contract**), although the average contract length was three-and-a-half years (CNUC 2010a).

Contract Structures

CNUC (2010a) and Goodfellow (1995) found three common contract types for utility vegetation management services: time and material (also referred to as time and equipment), unit price, and lump sum (fixed-price). Tate (1986) described the same strategies for municipal tree maintenance. Orr (2007) identified performance-based contracts as another strategy, which is a time and material contract modification.

In CNUC's survey, time and material was the most common vegetation management services contract structure in 2006, having been used by 53 percent of responding utilities, although that declined to 44 percent in 2012 (CNUC 2015). Unit price was reported by 22 percent in 2006 and 33 percent in 2012, making it the second most frequently reported contract strategy. Lump sum (fixed price) was third, having been adopted by 17 percent of respondents in 2006 (CNUC 2010a) and 19 percent in 2012 (CNUC 2015). CNUC noted a trend away from time and material toward unit price since 2002, when 70 percent of responding utilities contracted on a time and material basis, and only 14 percent by unit price (Figure 2.6).

Pre-inspection contracts are more heavily weighted toward time and material structures. Eighty-four percent of pre-inspection contracts were awarded on a time and material basis, with the remaining 16 percent split evenly between unit price and lump sum.

There are advantages and disadvantages to each type of contract structure. Selecting the type best suited to a system depends on conditions specific to a project or program, as well as management preference. Table 2.1 provides the relative risks and benefits of various contract types.

Time and Material Contracts

Time and material (T&M) **contracts** are based on hourly rates for labor and equipment (Tate 1986),

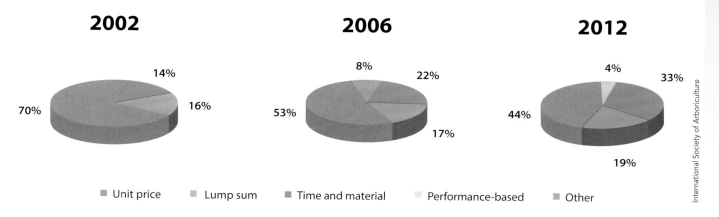

Figure 2.6 Percent of responding utility vegetation management programs with time and material, unit price, lump sum, and performance-based (incentivized) contract structures. *(Data from CNUC 2015)*

and payment is for time worked (Orr 2007). Tate (1986) considers time and material contracts to provide the best value for the contracting body, presumably since contractors know what they will make on a job, and don't have to pad their bids to ensure their expenses and profit are covered. Lough (1991) preferred time and material over unit price contracts for urban forestry services in New York, U.S. He found time and material contracts allowed the city greater management flexibility, reduced supervisory requirements, provided stronger control, and lowered costs. Orr (2007) observed that time and material contracts are low risk for contractors, but present higher risk for utilities, because utilities are responsible for productivity, performance indicators, best practices, planning, and to a lesser extent, work quality and customer satisfaction (Table 2.1).

Cieslewicz and Novembri (2004) found contrasting opinions among vendors on time and material contracts. Positive comments indicated that vendors considered time and material contracts to be nearly always the most economical contracting method for utilities. They pointed out that time and material contracts provide the most flexibility and stable workforce. On the other hand, cited disadvantages include low contractor financial return, the risk of complacency on the part of tree crews and contractor management, and greater demand on utility oversight to evaluate performance.

Performance-based contracts are time and material agreements modified to promote productivity and work quality (Orr 2007). The strategy attempts to draw positive elements from both time and material and unit price contracts. Orr (2007) promotes performance contracts as beneficial for both contractors and utilities as they share the risks and responsibility for productivity, quality, performance indicators, best practices, and other factors between the contractor and utility (Table 2.1).

Key performance indicators can include unit price components (e.g., cost to work a tree, acre [hectare], or mile [km]), as well as other elements like work quality, customer satisfaction, reliability, safety, International Society of Arboriculture certifications, or other special training. In this structure, contractors can receive a predetermined financial incentive for exceeding the key performance indicators. Some performance contracts also have penalty provisions if goals and objectives are not met.

Unit Price Contracts

Unit price contracts are billed for each unit worked (Tate 1986). Units can be trees, miles (km), acres (hectares), or some other specified work product (Orr 2007). Separate billing rates are sometimes established for various unit classifications, like different diameter size classes (Tate 1986). Orr considers contractor risk for unit price to be higher than that for time and material. Unit price

Table 2.1 Contract risks and benefits.				
Characteristics	Time and material	Unit price	Lump sum	Performance
Contractor risk	Low	High	High	Shared
Utility risk	High	Variable	Variable	Shared
Concern for productivity	Utility	Contractor	Contractor	Shared
Concern for quality work	Variable	Variable	Variable	Shared
Pay based on	Time	Work	Work	Work
Quality tied to payment	Sometimes	Sometimes	Sometimes	Yes
Safety	Contractor	Contractor	Contractor	Shared
Customer satisfaction	Shared	Utility	Utility	Shared
Utility reliability	Utility	Utility	Utility	Shared
Performance indicators	Utility	Shared	Shared	Shared
Storm response	Utility	Contractor	Contractor	Shared
Best practices/specialized equipment	Utility	Contractor	Contractor	Shared
Work planning	Utility/3rd party	Contractor	Contractor	Shared
Workforce stability	Shared	Utility	Utility	Shared
Work schedule	Utility	Contractor	Contractor	Shared
Cost management	Shared	Contractor	Contractor	Shared

Adapted from Orr 2007.

contracts are gaining popularity because they shift the onus for better productivity, best practices, work planning, and scheduling from the utility to the contractor (Table 2.1).

Some utilities have enjoyed improved production with unit price. Bell (2004) found the cost of trees worked on a unit basis in New Mexico, U.S. to be half that of those completed under time and material contracts. He reasons that the improvement is because contractors are motivated by market forces to improve productivity with unit price, but not for time and material structures. Tate (1986), on the other hand, considers unit price to be inferior to time and material contracts because trees are not uniform units, and the type of work differs substantially depending on species, size, location, and accessibility. He reasoned that2 these factors create unknown risks for the contractor that can lead to inflated bids.

Vendors commented on their ability to leverage their expertise and flexibility in deploying their resources as advantages derived from the inherent incentives and penalties in unit price contracts. On the other hand, they were apprehensive about the risks of limited profitability due to difficulty in accurate bidding, dependency on vendor trustworthiness, and increased utility responsibility for post auditing. They also expressed concern over lost focus on safety as they concentrate on production, greater risk of contractor default, the difficulty of work, and challenges to contractor-utility relationships if units are inadequately described (Cieslewicz and Novembri 2004).

Bell (2004) advises that utilities planning unit price contracts produce highly refined specifications. Potentially ambiguous terms, including seemingly obvious particulars—such as pruning, clearance, overhang and hazard trees—should be clearly defined. Utilities need to meticulously explain their concept of a unit and what work is expected in each case. For example, Public Service of New Mexico used separate rates to prune trees by lift truck and climb crews, but did not differentiate trees by size class, species, or any other distinguishing factor. Other utilities apply only one price to prune a tree, but might have different rate structures depending on geographic location (S. Tankersley, personal communication). Still others may break trees into size classes to set prices for pruning and removal. Public Service of New Mexico differentiated six distinct removal classes, each separated into lift or climb crews (Bell 2004). Successful unit price programs have also used unit classes for removals, but applied time and material rates for large (e.g., 36-inch [1-m] diameter) or difficult removals (like those requiring substantial roping and rigging (S. Tankersley, personal communication).

The contract should also be clear on specification clearance expectations (e.g., established distances or clearances that hold for a specified duration), conformity to pertinent ANSI A300 Standards and ISA's Best Management Practices series recommendations, or other requirements. The point is that each unit price program needs to provide contractors with clear expectations for work necessary on its particular contract (Bell 2004).

Utilities with unit price structures should have detailed workload assessments, so both the utility and contractors understand what needs to be done (Bell 2004). Many utilities with unit price contracts employ pre-inspectors. Unit price contracts may also require post-inspections or audits to ensure reported work is completed to specifications (Bell 2004). While it is the utility's responsibility to manage the contract and monitor the results, the contractors have to be given room to manage their operations (Bell 2004; Orr 2007).

Bell (2004) advises companies that bid on unit price contracts to be realistic, as underbidding in an effort to win an account can be financially disastrous. He suggests contractors fully understand the productivity of their crews in a particular area before bidding a unit price contract.

Lump Sum (Fixed-Price) Contracts

Lump sum (also called fixed-price) **contracts** are awarded on the basis of the price of an entire job (Tate 1986), with pay based on work (Orr 2007). A job could refer to an entire system or to a specific project within a system. Orr (2007) considers vendor risk with lump sum contracts to be higher than that for time and material contracts (Table 2.1). Lump sum contracts are similar to unit price contracts insofar as they shift responsibility for productivity, best practices, work planning, and scheduling from the utility to the contractor (Orr 2007). Lump sum contracts also place the burden of understanding workload on the utility. Fixed-price contracts have commonality with unit price contracts from a utility perspective insofar as lump sum contracts require refined specifications, as well as robust quality control and quality assurance programs.

Goodfellow (1985) advises that the key to successful lump sum contracting is detailed project specifications. He itemizes six critical sections that are important to lump sum bidding (Table 2.2). After a fixed-price contract has been awarded, scheduling, staffing, equipping, and supervising the work is the contractor's responsibility (Goodfellow 1985).

Goodfellow (1985) estimated that Wisconsin Public Service realized a 15 to 20 percent savings with fixed-price over time and material contracts. He observed that contractors benefitted because fixed-price contracts allowed them flexibility to maximize efficiency in applying human and equipment resources to the project. Wisconsin Public Service also enjoyed improved productivity as a result of increased contractor use of specialized equipment, creative crew composition, and application of incentive payments to crew members.

Some disadvantages of fixed-price contracts are the potential for contractors to take shortcuts that reduce customer goodwill and to compromise work specifications. Goodfellow (1985) advises that utilities increase monitoring to limit the potential damage from these two considerations.

Cieslewicz and Novembri's vendor survey (2004) related positive comments for lump sum contracts in providing the potential for significantly higher profits, control over the job, more stable utility budgeting, less supervision on the part of the utility, greater contractor flexibility, and increased opportunity for contractor creativity. On the negative side, vendors cited the potential for losses, emphasis on production detracting from

Table 2.2 Specification sections important to lump sum bidding.

Specification section	Description
Scope	• Provide written instructions delineated on maps. • Identify rights-of-way locations. • Describe all maintenance work within the project boundary. • Make work requirements as uniform as possible. Consider breaking areas such as rural and urban locations into separate units. • Create small work packages. Too large a project can restrict the number of bidders due to cash flow concerns, mobilization or demobilization costs, or increased labor and equipment demands. Too small a project can render it unattractive or unprofitable for potential vendors.
Notification	• Clearly establish responsibility for and characteristics of property owner notification (letter, personal contact, door hanger, phone call, etc.).
Clearance	• Objectively define requirements. • Start with concrete clearance limits and add modifiers. For example, less clearance may be allowed for major stems and branches or slow-growing trees. • Describe directional pruning. • Establish protocols for variation from required clearance requirements.
Slash disposal	• Chip and remove debris from urban sites, leaving firewood by request. • In rural areas, lop and scatter or blow chips, except in landscaped areas or where the property owner objects. It is the contractors' responsibility to obtain approval for blowing chips or lopping and scattering. If they fail to do so, it is their responsibility to make it right with the homeowner.
Reporting	• Project start and completion dates. • Establish requirements for record keeping and how often contractor contracts are required with the company. • Determine the frequency of invoicing and payment (such as percent completed). The final invoice should not be paid until the entire job is reviewed and accepted by the utility.
Exceptions	• Establish hourly time and material rates for unique situations (such as costly removals or storm response work).

Adapted from Goodfellow 1985.

training and safety, scope creep imposed by the utility on one hand and shortcuts on specifications by the contractor on the other, high start-up costs, risks to customer relations, and other concerns.

Procurement

Procurement departments are responsible for securing contracts. Since so many utilities contract for vegetation management services, most utility arborists—whether they work for utilities or contractors—will need to work with corporate procurement departments. Luefschuetz (2010) observed a trend in procurement at Fortune 500 companies toward identifying preferred vendors through competitive requests, increasing reliance on comprehensive use of master services agreements, implementing vendor performance management systems, carrying out meticulous purchasing processes, and increasing emphasis on aggressive negotiating. He cautions that while relationships are important, they are becoming less so as there has been a greater emphasis on pricing value to the contracting body.

Luefschuetz (2010) summarized large corporation procurement departments' agendas in contracting consulting, technology, and outsourcing services as:

- Optimizing expenditures and leveraging best practices across their organization
- Engaging fewer and more strategically selected vendors offering an array of expertise
- Preferring vendors willing to share risk and reward
- Favoring aggressive pricing, with volume and other discounts in exchange for privileged bidding status
- Promoting pricing that is consistent with the nature of subject services
- Championing best value—timely, high quality, and cost-effective delivery
- Encouraging healthy competition among vendors
- Stressing attention to compliance
- Advocating efficiency gains in time, costs, and control using e-procurement solutions
- Endorsing performance management systems under which vendors are evaluated against quality, time, and cost key performance indicators
- Recommending business and legal terms and conditions that offer the contracting body satisfactory risk and reward

Requests for proposals (RFPs) involve inviting pre-qualified vendors to bid on a contract. Increasingly these requests are made over the Internet (Luefschuetz 2010).

Cieslewicz and Novembri (2004) recommend that utilities promote competition to best serve their long-term interests. They make the following suggestions:

- Competitively bid vegetation management work.
- Grant awards to multiple vendors on any system that is large enough to accommodate them.
- Work to expand the pre-qualified list of vendors.
- Seek to cultivate new market players.
- Understand the true cost of the buying decision (note that the lowest hourly cost does not necessarily translate to the best value).

FLEET MANAGEMENT

Contractors are more likely to have fleet management responsibilities than utility vegetation management organizations. Among those utility programs that have "in house" fleets, larger companies are likely to have separate departments with responsibility over vehicles).

Fleet managers are often responsible for motorized equipment, or in some cases, any equipment requiring a license. Successful management requires knowledge of purchasing, accounting, shop management, corporate management, budgeting, financial analysis, data processing, statistics, safety, labor relations, negotiations, personnel administration, vehicle disposition, laws at a variety of levels, insurance, and taxes (Dolce 1994). Vehicle maintenance managers must have four priorities for fleet efficiency: durability, reliability, operating costs, and initial investment. Decisions should be based on data collected on the actual cost per mile to run a vehicle (Dolce 1994).

Evaluating Needs

The number of vehicles should be optimized to meet the organization's requirements. On one hand, employees need the resources to do their jobs. On the other, vehicles are expensive, and idle transport is wasteful and detrimental to the bottom line. So an efficient organization has sufficient inventory to operate smoothly without an unnecessary number of unused vehicles.

Fleet managers should begin with a vehicle inventory and encourage feedback from their staff on their needs. Underutilized vehicles should be eliminated and new ones added only after existing vehicles are being used effectively. Dolce (1994) recommends that fleet managers establish a systematic approach to adding vehicles, with an objective rational for each request. He considers the economic life to be an essential consideration to equipment managers. It requires a comprehensive consideration of both capital and operational unit expenses over time. Total expenses include:

- Depreciation
- Operations
- Maintenance
- Downtime
- Obsolescence
- Operator training
- Costs of carrying parts for it in inventory (storage, insurance, etc., but not the actual parts costs)
- Interest
- Alternative capital value
- Inflation

The best way to control costs is to replace vehicles and equipment before they are worn out. Therefore, vehicle replacement decisions should be made with the idea of attempting to find the sweet spot where vehicles are used long enough to depreciate their initial cost, but not so long that they need major repairs to keep them operating. If an existing vehicle costs less to operate than a replacement, it should be kept; if not, it should be replaced. Costs such as principal, interest, labor, parts, tax, registration, and fuel use of new versus existing vehicles along with the resale or salvage value of those due to be replaced are also important factors for consideration (Dolce 1994).

Matheny and Clark (2008) suggest life-cycle costing for fleet management, starting with a needs assessment accomplished by answering the following questions:

What is in the inventory (number, type, age, condition)?

What function does each piece of equipment perform? Is each needed?

What is the monthly mileage or hourly usage?

What is the life expectancy of each?

They recommend using the information to ensure that proper inventory of necessary equipment is available, and that there is a schedule established for replacement.

Maintenance

With increasing cost of equipment, preventative maintenance is becoming more and more important. Dolce (1994) advises businesses to take a long-term perspective and resist the temptation to reduce maintenance dollars in a misguided attempt to the save money.

A preventative program should include maintenance tasks that need to be performed periodically at set time or mileage intervals. Dolce (1984) recommends four steps in developing an affordable preventative maintenance program:

1. Specify preventative tasks necessary for each class of equipment. Evaluate equipment manufacturers' servicing and repair schedules.

2. Define responsibilities for performing maintenance tasks and assign them to:
   - Operators
   - Field mechanics
   - Local shops
   - Central shops
   - Vendor shops

3. Analyze the cost of the program. Identify the types of repairs (and vehicle classes) that demand the most labor and parts costs. Determine which tasks or repairs account for the majority of shop time. Target preventative maintenance to reduce high-cost repairs, particularly if they are repetitive for a particular type of vehicle.

4. Track costs for the preventative maintenance program and the costs of performing nonscheduled maintenance. Look for the highest volume or most expensive repairs and establish a plan to reduce the amount and expense of these tasks. The preventative maintenance program might have to be modified or other tactics required. For example, high volume costs for a given component might indicate a design flaw that is better addressed through equipment modification or procurement than preventative

maintenance. Moreover, large volumes of repetitive repairs might also indicate a need for targeted mechanic training. Finally, frequent breakdowns on individual units may be caused by faulty operator practices or abuse, which are best solved by operator training or stricter supervision.

GPS Vehicle Tracking

Increasingly, utility companies and contractors are utilizing GPS-based fleet tracking, also called **vehicle monitoring systems** (**VMSs**). These systems have internet connectivity capable of displaying a vehicle's location on a computer at the moment of observation. It can also provide movement history for the day or whatever period of time a manager desires (Figure 2.7). Moreover, a VMS can provide information on the speed of travel, whether or not a vehicle is operating at its current location, when a truck left or arrived at its base location, and other matters of consequence.

ACRT, a utility vegetation management consulting firm, tested a GPS tracking system on 10 percent of its fleet over a six-month period. They enjoyed a 10 percent fuel cost reduction due to decreased fuel use and fewer miles driven, as well as lower maintenance costs. At the same time, ACRT's at-fault accidents dropped 42 percent (Ross 2011).

PERSONNEL MANAGEMENT

The utility vegetation management labor force works for utilities, contractors, and consultancies. Regardless of whom people work for, human nature remains constant. How an employer treats their employees is critical to a program or company's success.

Sound personnel management is essential to develop highly functional teams. Michael Abrashoff (2002) observes that low pay is only the fifth most frequent reason people leave the military or any other job; the most common is disrespectful treatment from superiors. Goleman, Boyatzis, and McKee (2002) point out that people join companies, but

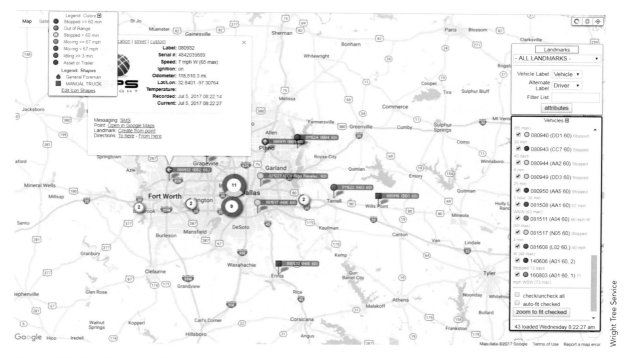

Figure 2.7 A vehicle monitoring system provides at-the-moment observation of vehicle movement. It enables supervisors to access crew locations as well as history, including speed of travel and arrival and departure times for a given location.

leave bosses. Supervisors with poor personnel management skills spread negative emotions, such as chronic anger, anxiety, or a sense of ineffectiveness. Negative emotions are bad for business because they distract workers' attention, disrupt positive performance, and ultimately drive the best and most ambitious employees away (Goleman et al. 2002).

Mindset

Dweck (2006) categorizes managers into two distinct mindsets—fixed and growth. Those with a fixed mindset consider an individual's qualities, like intelligence, moral character, or other talents to be inherited in the same manner as height or gender, and believe there isn't much anyone can do to improve them. She observes that a fixed perspective considers effort to be contemptible because it exposes a lack of talent. According to Dweck, supervisors with fixed mindsets feel they have risen to their positions because they are innately superior to others, and they use their power to validate and solidify their authority at the expense of their reports. What's more, people with fixed mindsets are compelled to repeatedly prove their superiority because those who are better at a particular skill represent threats against which there is no other defense. They are adversarial, mistrustful, and prone to micromanage their staffs. In so doing, they undermine the workplace by creating negative emotions that disrupt morale and productivity.

People with a growth mindset, on the other hand, consider qualities to be subject to improvement through hard work and practice, regardless of innate abilities. Effort is valued, as it invariably makes people better at what they do. Supervisors with a growth mindset are mentors. They encourage their employees' development and view the success of someone who reports to them as a personal achievement, even if a former employee advances beyond them in their careers. Managers with growth mindsets build the most functional and productive teams (Dweck 2006).

Emotional Intelligence

Goleman, Boyatzis, and McKee (2002) promote a management style centered on what they call emotional intelligence. From their perspective, emotionally intelligent management leads to resonant leadership, which creates the most productive teams. They promote what in all likelihood Dweck (2006) would consider a growth mindset insofar as emotional intelligence involves mentoring, and the resulting positive leadership skills can be learned and improved through practice.

Emotional intelligence consists of four domains: self-awareness, self-management, social awareness, and relationship management. Self-awareness requires an understanding and acknowledgement of one's core values, goals, strengths, and weaknesses. Self-management is derived from self-awareness and demands dedication to one's desired goals. In that regard, good leaders need to keep their emotions in check. Those who freely vent their anger, complain incessantly, or otherwise lose control of their negative emotions create an environment that distracts and demoralizes their employees. Bosses with poor self-management skills can't provide positive leadership. On the other hand, leaders who can stay positive and cheerful, even under pressure, create the resonance needed for highly productive teams (Goleman et al. 2002).

Social awareness requires empathy, which includes active listening and considering others' perspectives, among other behaviors. It enables leaders to take the most appropriate action for the benefit of the group. Relationship management includes collaboration, conflict management, and persuasion. Skilled relationship management is centered on consensus building.

Abraschoff (2002) provides a description of resonant leadership that he used successfully as captain of a naval vessel. He recommends the following management techniques:

- Lead by example
- Listen aggressively
- Communicate purpose and meaning

- Create a climate of trust
- Look for results, not salutes
- Take calculated risks
- Build up your people
- Generate unity
- Improve your people's quality of life

Motivating People

In his book, *SuperMotivation*, Dean Spitzer explains ways to energize employees and foster a high-functioning team. He discourages demotivating behavior that undermines the workplace, and advocates motivational techniques (Table 2.3).

Bennis (2009) asserts that trust is essential to gain and keep people's allegiance. He thinks that trusted leaders have four characteristics in common:

- **Constancy.** Leaders stay on course, they don't surprise the group.
- **Congruity.** Leaders are consistent in practice and philosophy. They "walk their talk."
- **Reliability.** Leaders provide support when it matters most.
- **Integrity.** Leaders honor their commitments.

Performance Monitoring and Evaluation

To develop and maintain highly functioning teams, managers have to monitor and evaluate employee performance. Matheny and Clark (2008) advise that reaching goals requires three steps:

1. Developing performance standards and measures
2. Measuring performance against developed standards
3. Adjusting performance to meet standards

Goleman et al. (2002) list what they consider to be key points in goal setting:

- Goals should be based on strengths, not weaknesses.
- Employees should have a central role in setting their own goals—they should not be imposed by someone else.
- Plans should be flexible and enable people to prepare for the future in a way that suits them. Autocratic planning imposed by an organization is often counterproductive.

Table 2.3 Motivators and demotivators.

Motivators	Demotivators
Add fun and variety to routine	Create a politically charged atmosphere
Provide employees with input and choice on how to do their work	Leave expectations unclear
Encourage responsibility and provide leadership opportunities	Create needless rules
Promote social interaction and teamwork among employees	Hold unproductive meetings
Tolerate learning errors and avoid harsh criticism	Promote internal competition among employees
Promote job ownership	Withhold information employees need to perform their work
Develop goals and challenges for all employees	Provide criticism instead of positive feedback
Provide ample encouragement	Tolerate poor performance so top performers feel taken advantage of
Make appreciation part of your repertoire	Treat employees unfairly
Develop measurements that show performance increases	Underutilize employee capacity

Source: Spitzer 1995.

- Plans must be realistic, with manageable steps: In all likelihood, plans that don't fit readily into a person's life and work will often be dropped within a few weeks or months.
- Plans that don't accommodate a person's learning style will demotivate them and lead to disinterest.

Goals should be based on **key performance indicators** (**KPIs**) (Matheny and Clark 2008). Ideally, key performance indicators should be **SMART**: specific, measurable, achievable, relevant, and timely (DelPo 2007). Matheny and Clark (2008) describe two levels of performance standards: programmatic and employee. Programmatic work objectives are KPIs that are achievable within a dedicated budget over a set time period, often a year. Employee goals establish expectations for an individual's work results within an established period.

Performance Appraisals

Performance appraisals should involve formal evaluations. Outlaw (1998) advocates evaluations in which supervisors not only explore their views of a team member's performance, but also invite the employee's own perceptions. He encourages managers to note a report's strengths, positive attributes, and desirable trends, then move on to focus on areas that need improvement.

Employee performance evaluations should ideally involve a dialogue between supervisors and their reports about work quality, problem solving, and suggestions for improvement. Matheny and Clark (2008) cite Graham and Hays (1986), who point out that errors in employee appraisal systems are demoralizing. Mistakes include gratuitous use of variables such as initiative, enthusiasm, and loyalty, which are not SMART. Rather, evaluations should be to the point, and spotlight performance without focusing on what might be perceived as personality flaws (Outlaw 1998). At the end of a performance review, both the supervisor and employee should set new goals and objectives, with the employee committing to continue successful qualities and improve on focus areas.

An employee's performance should not be a surprise to them at a once-a-year evaluation. Rather, supervisors should provide continual coaching and feedback regarding employee performance. Successful managers look for teachable moments. Wenderlich (1997) advises managers to praise in public, but criticize in private. He cautions that it's not enough to have an open door policy. Rather, supervisors should "manage by walking around," spending time in the field, talking to people who are working for them, and let employees know that they are free to talk with them about whatever is on their minds.

Labor Unions

A **labor union** is an organization of employees that collectively negotiates wages, benefits, working conditions, and other matters (Matheny and Clark 2008). A group of workers represented by a union local is a *bargaining unit*.

In the United States, the primary law governing collective bargaining is the National Labor Relations Act or NLRA (Schwartz 2006). The NLRA provides employees with the right to join trade unions and requires both sides to bargain in good faith (US Legal n.d.).

Weingarten Rights

Collective bargaining agreements often determine how employee performance is evaluated. In the United States, union members have a right to representation during interviews that will lead to discipline or that an employee has reason to believe will do so (Schwartz 2006). These rights were established by the United States Supreme Court in 1975 in the case of *National Labor Relations Board v. J. Weingarten, Inc.* (Schwartz 2006). The judgment produced the following rules, which are called Weingarten Rights:

- A union member may request representation before or at any time during an interview that will lead to discipline or that an employee has reason to believe will do so.
- In response to the request, the employer has three choices:

- Grant the request and delay questioning until a representative arrives.
- Deny the request and end the interview immediately.
- Offer the employee the choice of continuing without representation or ending the interview.
- If the employer denies the request for union representation and continues the meeting, the employee may refuse to answer questions.

Employers are not obligated to inform workers of their Weingarten Rights.

Representation is provided by union stewards. At disciplinary meetings, employers are obligated to allow the steward to advise and assist the employee in presenting the facts in the following ways (Schwartz 2006):

- The supervisor or manager must inform the steward of the nature of the investigation.
- The steward and employee have the right to a private meeting before questioning begins.
- The steward may speak during an interview, but not end the meeting.
- The steward may object to a confusing question and ask for clarification.
- The steward may advise the employee against answering abusive, misleading, badgering, or harassing questions.
- After questioning, the steward may give arguments to justify the employee's conduct.

Grievances

Grievances are a type of dispute resolution used by companies to address complaints by employees, suppliers, customers, or competitors. Grievances are often used to resolve disagreements between labor unions and employers. Grievances involve a formal process, assign responsibility to specific individuals at each stage, establish timelines, and detail the documentation required as part of the process (US Legal n.d.). The grievance process is typically composed of several escalating steps, but the objective is to resolve the disagreement at the lowest level possible (Spalding n.d.).

Grievance mediation is a voluntary, non-binding procedure using neutral labor mediators to help both parties negotiate a satisfactory resolution to a labor contract disagreement. Many states and provinces have labor mediators available (Spalding n.d.; Newfoundland 2011). Grievance mediators often meet with disputing parties separately to better understand how the disagreement degraded to the point where it required mediation.

Spalding (n.d.) describes the advantages of mediation over standard procedures as being:

- More expedient
- More cost-effective
- Less contentious
- More cooperative
- Independent of a contract remedy
- Capable of resolving the real issue

Of these, perhaps the most important benefit of mediation is that it is independent of a contract remedy because the source of a dispute may be outside of existing contract language. This allows a mediator to attempt to get to the bottom of a dispute and work to resolve the underlying issue, rather than just the immediate grievance. Moreover, if the disagreement results from one side's misinterpretation of contract language, the mediator may be able to persuade the mistaken party that their position is without merit, and that continuing to pursue their case will only create unnecessary expense and enmity (Spalding n.d.). Ultimately, successful mediation is dependent on a desire of both parties to cooperate toward a mutually agreeable resolution.

Many labor contracts have a binding arbitration provision as the final step of the process. In binding arbitration, two sides agree on an arbitrator and to be bound by that arbitrator's decision. An arbitration is a formal proceeding with the final outcome dependent on the veracity of both parties (Spalding n.d.).

Right to Work

Right-to-work laws prohibit mandatory union membership as a condition of employment (National Right to Work Legal Defense and Education

Foundation n.d.). Twenty-eight states in the United States have right-to-work laws. Proponents of the right to work argue that people should be free to choose whether or not they want to join a labor union. They further maintain that voluntary membership strengthens an organization over what it would be if it was comprised of unwilling members (Rodriguez 2017). Others argue that right to work is a deliberate attempt to weaken unions and consider those who refuse membership as freeloaders who want to enjoy the advantages of the increased wages and improved working conditions that unions provide without paying to support the cost of obtaining those benefits (Johnson 2005).

SUMMARY

Utility vegetation management is a multifaceted and often complex enterprise. It requires effective program planning, an understanding of the environment and effective control measures, work scheduling, managing large sums of money, program management, knowledge of computers, contracting, and personnel management. While this chapter covered each of these topics, readers should be aware that every topic is more involved than can be covered in any single chapter. Those wanting more in-depth study are directed to the references section of this book.

CHAPTER 2 WORKBOOK

Fill in the Blank

1. Communication between vegetation managers, contractors, supervisors, and workers should be both _____ and _____.

2. Long-term planning is _____, while short-term planning is _____.

3. _____ are documents that describe, instruct, and direct work results.

4. Research has shown that trees are more likely to cause outages on _____-phase lines than on _____-phase lines.

5. A results-based strategy designed to target trees as close as possible to the time that they interfere with facilities is known as _____-___-_____ management.

6. A _____ _____ structure is a prioritized outline of job components that must be completed for a successful project outcome.

7. For most utilities, vegetation management departments are considered _____ _____ that do not generate income.

8. Purchases of equipment or line construction are considered _____ expenses, whereas supply purchases or salaries are considered _____ expenses.

9. Time and material contracts, which are based on hourly rates for labor and equipment, are generally _____ risk for contractors and _____ risk for utilities.

10. _____ _____ contracts are gaining popularity because the responsibility for productivity, work planning, and scheduling rests with the contractor.

11. Depreciation can be either _____-____, in which costs are spread out over the lifetime of an item, or _____, in which the amount of depreciation taken is higher immediately after the purchase of an item.

Multiple Choice

1. All of the following are approaches to work planning, *except*
   a. cycle-based
   b. results-based
   c. objective-based
   d. crisis management

2. SWOT analysis is a strategic planning tool and an acronym, meaning
   a. Strengths, Weaknesses, Opportunities, Threats
   b. Skill, Work, Objectives, Tactics
   c. Scope, Wants, Observation, Talent
   d. Safety, Weaknesses, Objectives, Targets

3. Relative to the cost of routine maintenance, reactive maintenance can cost
   a. about the same
   b. one and half times as much
   c. twice as much
   d. five times as much

4. Which of the following is not a part of the constraint triangle?
   a. time
   b. skill
   c. cost
   d. scope

5. With respect to personnel management, what is the most common reason that people leave jobs?
   a. low pay
   b. workload
   c. relationships with coworkers
   d. treatment from supervisors

CHALLENGE QUESTIONS

1. Describe how a crisis management approach to utility work impacts work crews, service reliability, and efficiency.

2. Explain the advantages and disadvantages of a circuit approach and a grid approach to vegetation management.

3. Describe the characteristics of an effective employee performance evaluation program.

4. Explain how a manager could use a geographic information system (GIS) to enhance productivity and efficiency in their utility vegetation management system.

5. Outline the steps recommended for effectively preparing a budget.

3

Utility Pruning

OBJECTIVES

- Recognize the purpose of utility pruning to both internal and external stakeholders.
- Discuss how trees affect the intended use of rights of way and easements.
- Describe how different pruning systems can be utilized to achieve pruning objectives.
- Decide appropriate pruning cuts and styles to achieve different management objectives.
- Identify the main objectives of utility pruning.
- Explain the importance of determining appropriate pruning intervals.
- Determine appropriate utility pruning methods for remote forested environments.

KEY TERMS

- branch bark ridge
- branch collar
- branch removal cut
- branch union
- chemical pruning
- codominant stem
- directional pruning
- easement
- epicormic shoot
- excurrent
- frond
- heading cut
- lion tailing
- maintenance interval
- mechanical pruning
- monocot
- natural pruning
- node
- pollarding
- preventive maintenance
- pruning system
- reduction cut
- response growth
- right-of-way (ROW)
- roundover
- service reliability
- shearing
- structural development
- subordination
- topiary
- topping
- tree growth regulator (TGR)
- urban forest
- utility facilities
- utility forest
- utility service
- whorl
- wound treatment

INTRODUCTION

Utility tree pruning plays an important role in the safe and reliable delivery of vital services, such as electricity and communications, and in minimizing interference with transportation corridors, sidewalks, street and traffic lighting, signage, buildings, pipelines, and other rights-of-way (ROWs) (Federal Highway Administration 2008). Utility pruning can reduce the likelihood of branch failure and impact to power lines, as well as the incidence of wildfire due to such contacts. Though tree and branch failure is frequently the cause of **utility service** interruptions, utility arborists must also recognize that trees and **urban forests** enhance quality of life and provide valuable services that are quantifiable, including some that directly benefit utility interests.

This chapter is written with the assumption that the reader has a general knowledge of arboriculture, including tree risk assessment, basic tree biology, and pruning principles as outlined in ISA's *Arborists' Certification Study Guide*, Best Management Practices series, and other publications.

Considering the importance of utility pruning, and the fact that much of the work is conducted on private property and is highly visible to the public, it is critical that utility arborists adhere to the highest professional standards. Utility pruning information provided in this chapter is based on published research and proven methodology as presented in current industry standards for pruning and tree risk assessment, and other sources. Application of these methods will ensure that the investment is returned to customers in the form of safe and reliable utility services, and healthy, well-maintained trees and urban forests.

Utility arborists should always bear in mind that trees are unique and dynamic organisms. Tree growth rates and stability are influenced by climate zone, annual moisture, soils, frequency and intensity of storms, length of growing season, elevation, terrain, species, pests and disease, and combinations of these and other factors. In addition, facility priority, accessibility, equipment availability, workforce capabilities, local restrictions, and other factors influence how the work can be specified and completed. All of this means that practitioners will be required to adapt and apply professional judgment on a continuing basis.

Given this variability in conditions, it is not reasonable to expect that all risk from trees can be eliminated. Instead, utility arborists should concentrate on managing risk by applying proven practices that systematically reduce incidents that could disrupt service (Utility Arborist Association 2009). It should also be understood that communities and members of the public value trees for a variety of reasons. Utility arborists must be prepared to listen to the concerns of various stakeholders, communicate the benefits of utility pruning, and adjust when necessary.

PURPOSE OF UTILITY PRUNING

Utility pruning is performed to maintain an acceptable level of safety for the public and for workers maintaining **utility facilities**, and to ensure **service reliability** by reducing the risk of interference from vegetation. If not properly maintained, vegetation may also damage infrastructure and impede access to utility facilities used by maintenance and repair personnel (ANSI 2017b).

Utilities operate and maintain overhead power lines in a variety of settings; some are nearly natural, such as unmaintained forests, but more often in environments that have been significantly manipulated by human activities, such as most urban settings. Regardless of whether they are part of natural, altered, or purely contrived landscapes, trees with the potential to affect the operation of utility facilities can be characterized as the **utility forest** (Goodfellow 2008).

Trees in the utility forest present multiple concerns to utility owners, property owners, and indeed, to all end users of utility services (Figure 3.1). Utility facilities can be affected by trees growing into lines from below or alongside, or by trees failing and striking lines from farther away. Risk from trees growing into lines is managed primarily by pruning to maintain clearance, whereas risk from trees that may fall in is managed through regular inspections, followed by pruning or removing trees as necessary to reduce risk of failure and impact.

In some jurisdictions, government authorities have mandated minimum clearances between energized conductors and surrounding vegetation, or have enacted other compulsory measures involving pruning frequency, customer communications, or adherence to published industry standards. Utility tree maintenance programs must be designed to comply with such requirements.

Industry standards and best practices recommend the use of **preventive maintenance**, often at planned intervals, to increase the efficiency and effectiveness of utility vegetation management programs. Inspection regimens, maintenance intervals, and the amount of pruning performed should be adjusted to accommodate line priority, tree growth rates, infestations, typical failure patterns, or other variables. Excessive pruning is needlessly costly, may unnecessarily injure trees, and often leads to customer relations concerns. At the same time, inadequate pruning may result in unacceptable risk of power outages, damage, personal injury, or compliance violations.

Utility pruning operations are performed on rights-of-way across private and public property. A **right-of-way** is land with use rights designated for an intended purpose, such as a utility line. The right-of-way may include portions of many individual properties through the use of legal **easements**. An easement authorizes one party to use the property of another for a specific purpose. Utility arborists should be familiar with the sizes and specific terms of rights-of-way and easements in the areas in which they work (Xcel Energy Inc. 2007).

Utility easements often include only a portion of an individual tree. However, trees, branches, or other parts may pose significant risk to utility facilities whether or not they are within an easement. When a worker observes such conditions that are beyond the scope of the work assignment, a supervisor should be informed of the situation (ANSI 2017b).

Figure 3.1 Tree failures have a significant effect on utility services.

Utility arborists should recognize that maintenance of the portion of the tree that is outside of the utility easement and has little or no likelihood of failure and impact to utility facilities is the responsibility of individual tree owners (Figure 3.2). Furthermore, utility arborists are not obligated to inspect the tree beyond the scope of their work assignment, nor are they required to inform tree owners of conditions they may observe. This limitation is due to the impracticality of informing large numbers of individual tree owners who may be difficult to locate. Of course this does not preclude an employee from answering customer questions or pointing out a problem if the opportunity arises (ANSI 2017b).

UTILITY PRUNING OVERVIEW

Standards and Best Practices

The pruning methods described in this chapter comply with accepted industry standards and best management practices developed by ISA and the Tree Care Industry Association (TCIA), as well as other peer-reviewed guides and texts. The ANSI A300 Standard defines terms and provides a basic framework for written specifications, training manuals, and other tree maintenance information. Best practices guides often expand on information provided in standards, and provide explanations and interpretations for end users. While it is important for practitioners to understand and apply standards and best practices as they perform their work, it is sometimes necessary to adjust to meet varying field conditions. Decisions to deviate from recommended standards and best practices should be undertaken only after careful consideration.

Pruning Systems

Achieving and maintaining pruning objectives requires a systematic approach. This implies returning to maintain trees at prescribed intervals or as needed. Various **pruning systems** are used to maintain plants in a desired form, to maintain

Figure 3.2 Utility arborists are responsible for only the portion of the tree that could affect the facility. Maintenance of the remainder of the tree is the responsibility of the tree owner.

good structure, enhance the production of fruit, or as part of other objectives. Some systems, such as topiary, may require frequent and intensive pruning throughout the course of every growing season, while others require less intervention. The systems mentioned below have all been used in some form in utility arboriculture. For more information on pruning systems, refer to ISA's *Best Management Practices: Tree Pruning*.

Natural

A **natural pruning** system is performed with an understanding of the tree's natural characteristics and growth patterns, while allowing for changes in appearance due to the need to accomplish pruning objectives. Most utility pruning falls within the natural system. While many would dispute that utility pruning leaves trees with a "natural appearance," tree form in any setting may be affected by site factors, such as the presence of other trees, or the need to maintain clearance from buildings or traffic. Utility pruning should be performed with an understanding of the natural shape and form of trees and their responses to pruning. The appearance of the tree following utility pruning will vary depending on individual circumstances.

Pollarding

Pollarded trees are maintained at a predetermined size. This system begins by making strategically placed heading cuts when a tree is young, followed by regular removal of all new shoots that arise from those points, without damaging the pollard heads (i.e., woody knobs) that develop at the sites of the original cuts (Figure 3.3). Annual removal of shoots is preferred; however, up to three years may be acceptable for some slower-growing species in cooler climates.

In the past, utility pruning was sometimes accomplished by "topping" or "rounding over" (described below). Such maintenance was often performed at intervals of five years or more, and with little attention to the placement of the cuts. Many people mistakenly refer to this now discredited practice, and indeed to any such indiscriminate practice, as pollarding.

Figure 3.3 Pollarding is a pruning system that requires regular, ongoing maintenance. Pollarding should not be confused with topping, which is an unacceptable pruning practice.

When performed properly, pollarding is very effective in maintaining the desired size and shape of suitable trees species. Pollarded trees effectively compartmentalize initial heading cuts and subsequent cuts to remove sprouts (Harris et al. 2004). However, the required frequency of pruning and accompanying cost usually make routine use of this method impractical along utility corridors.

Topiary

Topiary maintains plants in a desired shape, such as hedges, which are woody plants grown closely in rows. Hedges in formal landscapes are usually maintained at frequent intervals with specialized hand tools. However, in rural areas where hedges form the boundaries of fields and property lines, or line the edges of roads and paths, they can grow to heights that affect the operation of overhead utility facilities. To prevent conflicts, hedges are often maintained by periodic **shearing** at intervals of one or more years, using either hand-operated tools or equipment mounted on machinery (Energy Networks Association 2007).

Topping

Also known as **roundover** or hat rack, **topping** is the discredited practice of stubbing all or part of the crown of a tree without regard for tree health or structure, the location of lateral branches, or the expected response of the tree (Figure 3.4). This once widespread practice is not technically a pruning system, as it is ineffective in achieving management objectives. Furthermore, it is unprofessional and is unacceptable because it severely damages trees and encourages rapid re-growth. Vigorous trees respond to topping with a flush of fast-growing sprouts, which can rapidly overtake conductors (Figure 3.5). Less vigorous trees may not survive the treatment (French and Appleton 2009). Repeated topping depletes stored nutrients, weakens even vigorous trees, and increases susceptibility to decay and other pathogens and pests. In contrast, a natural system minimizes injury, makes only prescriptive cuts, and directs future growth away from utility facilities.

Figure 3.4 Topping is not an acceptable pruning practice.

Figure 3.5 Topping results in fast-growing sprouts and weakly attached branches that pose greater risk.

Pruning Cuts

The quality of pruning cuts has a direct impact on the overall effectiveness of utility pruning. Poorly made cuts may introduce decay or promote unwanted growth, potentially increasing risk of failure and future costs. Relatively small cuts compartmentalize better, close more rapidly, and have less impact on overall tree health (Gilman et al. 2013). In general, the need to make large pruning cuts will be reduced when pruning is performed more frequently and growth is directed away from facilities.

Pruning operations should not result in damage to other parts of the tree, other trees, or surrounding property. When making large cuts, or when wood splitting or bark tearing is likely, branches should be pre-cut. Whenever necessary, large branches should be carefully lowered to the ground (ANSI 2017b).

Figure 3.6 shows several examples of pruning cuts. A **branch removal cut** is made when removing the smaller of two branches at a union. The cut should be made close to the parent stem and without leaving a stub. When making the cut, the **branch bark ridge**, an area of bark that forms at the union between the branch and the parent stem, and the **branch collar**, the area of transition at the base of the branch, must be left intact. This is also true when removing dead branches, in which case the branch collar may extend some distance from the parent stem.

When cutting from the outside, as when removing a branch with a narrow angle of attachment, it is also important to avoid damaging the branch collar, and to avoid accidentally cutting into the remaining stem (ANSI 2017b).

A **reduction cut** removes the larger of two branches, or a **codominant stem**, at a union. When

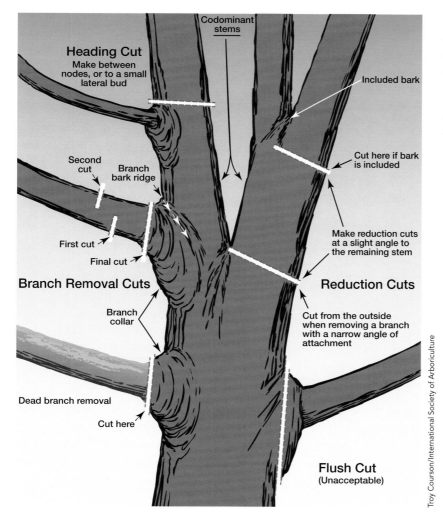

Figure 3.6 Examples of proper execution of branch removal cuts, reduction cuts, and heading cuts. A method of precutting large branches is also shown. Flush cuts, in which the branch collar is removed, are considered unacceptable.

making a reduction cut, the cut should be made at a slight angle (Figure 3.6) (ANSI 2017b).

A **heading cut** removes a branch or leader either between **nodes**, or to a much smaller lateral branch, or to a bud (Harris et al. 2004). The use of heading cuts should be reserved for certain situations, such as subordination of a leader, or when a heading cut may be preferred over making a larger wound elsewhere in the tree. When making heading or reduction cuts, the remaining lateral branch, or sprouts that arise from it, must be sufficient to sustain the branch (ANSI 2017b).

Completed cuts should be flat and even. The bark surrounding the wound should not be loosened or otherwise damaged. Wounds should close evenly from all sides (Figure 3.7). It is not necessary to use **wound treatments**, unless in response to a specific threat against which such treatments are known to be effective—for example, when pruning oak trees when and where they are susceptible to oak wilt (ANSI 2017b; O'Brien et al. 2011).

Figure 3.7 Pruning wounds should close evenly from all sides.

Tree Response to Utility Pruning

Utility arborists should be familiar with the characteristics of trees in the areas where they work, and plan according to how those trees are likely to respond to pruning—including the amount, type, and rate of growth. Factors that influence tree response to pruning include:

Species. Some trees, such as sycamore (*Platanus* spp.), cottonwood (*Populus* spp.), eucalyptus, willow (*Salix*), and others are known for prolific sprouting from lateral and latent buds following pruning. Other trees sprout less vigorously (e.g., certain oaks and many slower-growing angiosperms) or not at all, like most conifers, which respond instead with more growth at branch tips (McConnell et al. 1998).

Location, size, and quality of pruning cuts. Branches pruned to parent or suitable lateral branches sprout less than those that have been rounded over, stubbed, or pruned to relatively small laterals. In general, trees produce more sprouts when larger branches are pruned back to smaller branches (French and Appleton 2009).

Amount removed and tree vitality. In general, the larger the percentage of crown removed, the stronger the expected growth response from a tree in good health, with plenty of stored reserves (Cline 1997). In contrast, removing a large proportion of the foliage of an unhealthy tree will further reduce health.

Site factors. Soil conditions, site disturbances, available moisture, competition from other plants, and other factors can affect the direction and intensity of growth following pruning.

Structural and Directional Pruning

Utility arborists should adjust the amount of pruning on each tree depending on structural characteristics, including tree age, typical species or cultivar form, mature size and shape, past pruning practices, and the expected growth response following pruning. For example, a tree with a spreading (decurrent) habit adjacent to overhead lines is likely to require more pruning than a similarly placed tree with an upright growth (excurrent) pattern (Figures 3.8a and b). Likewise, a spreading tree growing directly beneath overhead lines can be directionally pruned more easily than one with an upright habit.

Past guidance has recommended the removal of entire branches that are growing toward facilities, rather than multiple small branches, resulting in fewer cuts overall (Shigo 1990; ANSI 2008). While this is sometimes the best approach, it may be counterproductive, especially if the objectives could be met while removing less living material from the tree. Less material removed reduces impact on tree health, the volume of debris that must be handled, and the overall amount of time required, especially if chipping and disposal time is included.

In addition, research has shown that differing lengths of branches and leaders result in varying frequencies of oscillation when trees are loaded by wind. These variations dampen the effects of wind load and reduce risk of failure (James 2003).

The complete removal of large branches may reduce the tree's ability to dampen loads. This can be avoided by making cuts to lateral or parent branches that pose little or no risk to the facility.

Avoiding complete branch removal also reduces the number of large pruning wounds, which may take years, if ever, to close, increasing the risk of decay in the parent stem. The health of the tree, and the long-term effect on growth rates, risks posed, and the amount of pruning that will be required in the future should all be considered when determining which branches to remove (Figure 3.9).

Rather than removing an entire large branch, the health and appearance of the tree may be preserved by subordinating or reducing selected branches. **Subordination** is a **structural development** pruning technique that uses reduction or heading cuts to reduce branch length relative to other similar branches. The subordinated branch grows at a comparatively slower rate, providing

Figure 3.8 A tree with a spreading habit (a) is likely to require more pruning than a tree with an upright growth pattern (b). Utility arborists should take into account tree form and expected tree growth rates when pruning.

Figure 3.9 Multiple factors, including tree species and health, risk posed, and future pruning requirements, should be considered when determining which branches to prune.

the option of removing the branch over more than one growing season or **maintenance interval** (Gilman et al. 2013). Subordination is used selectively on specific branches to improve structure, and should not be confused with topping or rounding over, which are performed without regard to tree structure, and are deleterious and discredited practices.

If the subordinated branch is retained, subsequent pruning should remove only vigorous shoots growing toward facilities, leaving those that pose little or no risk. Allowing healthy, non-interfering branches to grow may result in the cleared utility space becoming a smaller proportion of the crown over time.

Trees that have been rounded over or topped in the past and subsequently converted to directional pruning may have extensive **epicormic shoots** on old, decayed stubs. As these trees grow larger in relation to utility facilities, risk of failure may increase due to decay and the added weight of the branch (Figure 3.10). Risk to utility facilities can be mitigated by decreasing the load on decayed **branch unions** through reduction of epicormic branches, or by selectively pruning them as necessary. Such trees may also be good candidates for removal.

UTILITY PRUNING OBJECTIVES

Risk Reduction

The majority of tree-related service interruptions are caused by trees and branches falling onto utility facilities, as opposed to growth from alongside or below. This implies that pruning for clearance alone has limited efficacy in improving service reliability (Rees et al. 1994; Goodfellow 2005). A strategy that focuses on risk reduction by pruning and removing trees with a higher likelihood of failure and impact to utility facilities has been shown to have a significant effect on improving the reliability of utility service (Simpson and Van

Figure 3.10 Past pruning practices can affect future risk posed by trees. This tree had been previously topped, resulting in epicormic growth on old stub cuts.

Bossuyt 1996). While it is not possible to anticipate or prevent every occurrence, trees and branches that pose greater risk can be systematically identified and the risk can be mitigated as necessary by reduction or removal (Goodfellow et al. 2009).

Branches with certain characteristics are more likely to fail than healthy branches with strong attachments (Dunster et al. 2017). Typically, these include branches that are:

- Dead or declining
- Broken or partially failed, due to an earlier injury
- Adventitious (arose as sprouts, e.g., from old pruning wounds)
- More upright or codominant, especially when included bark is present at the branch union
- Overextended or poorly tapered, especially with heavy loads toward the distal end (e.g., **lion tailing**)
- Showing significant defects that compound any of the above, such as large cavities or decay

It should be understood that the presence of a defect alone may not indicate a significant structural weakness. Trees compensate for structural defects with **response growth**, which should be evaluated as part of a risk assessment process. In addition, some species are known to exhibit certain failure patterns, or be more prone to failure overall (Smiley et al. 2017). Utility arborists should take these factors into account when writing pruning specifications. For more information on the process of tree risk assessment, refer to Chapter 4 in this text, as well as the ISA *Tree Risk Assessment Manual*.

Clearance

Maintaining clearance allows for visual inspection of facilities, improves accessibility for maintenance personnel, and reduces risk of accidental line contact for workers and the public. Clearance also reduces incidental tree–line contact, which can affect power quality, and may increase risk of fire. While these benefits are important, it should be understood that trees and branches falling from beyond specified clearance distances have a greater impact on service reliability of distribution systems than trees growing into and contacting power lines (Goodfellow 2005).

Minimum clearance distance between tree branches and utility facilities is sometimes specified, either due to regulatory requirements or individual utility policies. When practical and unless expressly prohibited by law, such pre-established clearing limits should allow exceptions for situations where obtaining less than minimum clearance poses little or no additional risk, or conversely, when obtaining required clearance actually increases risk, either by causing vigorous re-growth or by excessively damaging a tree (Odom 2010).

When adhering to clearance requirements, a proper pruning cut should be made beyond the specified distance whenever practical (Figure 3.11); however, arborists should consider the effect of large pruning cuts on the structural integrity of the tree.

Determining Amount of Clearance

The amount and extent of pruning that is required in any given setting will vary depending on many factors, including:

- Priority of the facility (e.g., line voltage, overcurrent protection, number and type of customers served, etc.)
- Facility construction (e.g., pole height, crossarms, vertical, bundled, tree wire, amount of conductor sag, etc.)
- Planned maintenance interval
- Position of the tree in relation to the utility facility
- Work specifications
- Typical weather patterns in the area
- Individual tree characteristics (e.g., species, condition, expected growth rate, wood strength, mature size, shape, expected response to pruning, and other factors)
- Site factors, such as trees, buildings, terrain, soil, and other features that contribute to the shape, growth rates, or stability of trees

Prioritizing Resources for Tree Risk Assessment and Mitigation

In risk assessment, the amount of risk is a function of the probability of the occurrence of an event and its consequences, sometimes expressed as a simple formula:

Risk = Probability x Consequences

Consequences are the result of an event, for example, the failure of a tree or part of a tree striking a target, such as a utility facility. In utility arboriculture, the consequences of a tree failure could be the loss of service to a single property, or—should a tree strike a critical transmission line—an entire city. Other consequences could be personal injury or property damage due to electrical contact, fire, or the cost of associated fines or claims.

Probability is the chance of an event occurring, expressed as a percentage. In tree risk assessment, it is not possible to gauge the probability of an individual tree failure with numerical accuracy. However, it is possible to identify conditions that may increase the likelihood of failure of one tree or branch over another, based on documented patterns of other similar failures and the results of controlled experiments (Goodfellow et al. 2009; Dahle 2006). Both the likelihood of failure and impact to a facility and the consequences of such an event vary widely, depending on a range of factors.

Tree contact with high-voltage electric transmission lines has resulted in severe consequences, such as widespread blackouts (Cieslewicz and Novembri 2004). In North America and Europe, regulatory agencies have mandatory compliance requirements for vegetation management (NERC 2008; European Commission Directorate-General for Energy and Transport 2006). These highest-priority facilities are inspected frequently, and in most cases, incompatible trees are removed rather than pruned.

The vast majority of electric utility pruning takes place on distribution systems, which generally have voltages ranging from 3 to 35 kV. Focusing efforts on risk reduction pruning can increase the effectiveness of the vegetation management spend; however, the consequences of tree failure and impact to distribution lines vary considerably, depending on location. The information used to assign a level of priority to a particular utility facility takes into account many factors, including:

- Type of facility (e.g., transmission, distribution, secondary, substation, critical hardware, etc.)
- Number of customers potentially affected
- Number of outages experienced in the past
- Financial considerations, including available funds for mitigation
- Construction factors (e.g., age, height, resiliency, location of fuses, etc.)
- Other site-specific or regional risk factors, such as wildfire
- Government regulatory requirements (e.g., NERC, state or provincial mandates, etc.)

Other factors may also play a role in prioritization. For example, an industrial complex that employs a large number of people might technically be a single customer. However, the consequences of a service interruption to that facility could affect an entire community.

Given the complexity of electric utility distribution systems, resource allocation for tree risk assessment and mitigation should be determined with input from specialists, such as reliability engineers, who have knowledge of where focused risk reduction efforts will have the greatest effect.

Figure 3.11 If a clearance distance is specified, such as 10 feet (3 m), the cut should be made at the next suitable branch union.

Considering these factors, relatively little work may be necessary on some trees, while much of the crown may require reduction or removal on others. Also, with so many variables, there may be multiple acceptable approaches, depending on the arborist's interpretation of the specification. Depending on local policies, the tree owner's preferences may also be accommodated to some extent. While every effort should be made to avoid unnecessary damage to trees, the stated objectives of the utility vegetation management program must be accommodated (ANSI 2017b).

It is important to inform tree owners that achieving required clearance objectives may result in considerable change in the appearance of their trees. Concerned tree owners should also be advised that the initial visual impact of utility pruning will be softened as trees respond with new growth. For more information on communicating with the public, see Chapter 7.

Maintaining Structure

Over time, repeated pruning and the choices utility arborists make about which branches to prune will affect the structure of trees. If informed choices are made, tree structure can be maintained and sometimes improved for the site, and both long-term maintenance costs and the risks posed by trees to utility facilities can be better managed.

Directional Pruning

Directional pruning is accomplished by reducing or removing interfering branches back to lateral branches or parent stems that are growing away from the facility (Figures 3.12a and b). These lateral branches should be of sufficient size to become dominant, thus discouraging the growth of sprouts. This method, (known in the past as "drop-crotching") may also be referred to as natural pruning (Blair 1940), or "natural target pruning" (Shigo 1990). Directional pruning is most effective when performed at frequent intervals and when tree characteristics, such as typical species size, shape, expected growth rate, and other factors are taken into consideration. In this way, tree architecture can be established to reduce risk to utility facilities, and to minimize the amount of pruning required in the future.

How directional pruning affects tree form depends on the tree's natural growth habit and where the tree is relative to the facility. Trees closer to facilities will necessarily be affected more. Trees growing directly beneath facilities assume a different form than trees growing beside them (Figures 3.13a and b). The results of directional pruning are sometimes criticized because tree form may not be symmetrical; however, tree crowns in many other settings, including natural landscapes, are often asymmetrical due to the presence of other trees, structures, available light and other site factors.

Tall trees with an **excurrent**, or upright, growth habit (typical of many coniferous species) that are growing directly beneath a facility are often not good candidates for directional pruning. To obtain the required clearance from these trees, it may be

Figure 3.12 Directional pruning encourages growth in a desired direction (a), whereas topping or rounding over often results in a flush of sprouts (b).

Figure 3.13 A tree growing directly beneath facilities (a) will be pruned to a different shape than a tree growing beside them (b).

necessary to make reduction cuts at small laterals, or at **whorls**. When practical, such trees should be removed and, when appropriate, replaced with compatible species.

Managing Overhang

Large trees with branches overhanging utility facilities should be managed to reduce the likelihood of branch failure. Whether to completely remove or to reduce and maintain overhanging branches depends on factors such as:

- Line priority
- Tree species and condition, including branch length and size
- Susceptibility to storms
- Community norms

Removing all overhang is costly, and is not always necessary. When appropriate, overhanging branches can be managed by removing dead and weakly attached branches, and reducing branch length. In addition to significantly reducing risk of branch failure, this practice minimizes stress on trees, and can ease property owner concerns by preserving the amenity value of trees (Figure 3.14).

PRUNING INTERVALS

Utility pruning should be performed often enough to prevent vegetation-related problems, but not so frequently that costs are unnecessarily driven up. When considering the cost of vegetation management, external factors, such as the cost of tree-caused outages to end users, should be a consideration.

When and how often utility pruning is performed depends on a variety of factors, including the species and condition of trees and the type of utility facility. However, the fact that trees are alive, growing, and responding to a constantly changing environment adds a level of complexity to this

Figure 3.14 Whether to reduce or completely remove overhang depends on target importance and tree condition.

equation that is not necessarily present when maintaining other kinds of infrastructure. For example, several years of excessive rainfall may increase growth rates, throwing off planned maintenance intervals. Likewise, a drought may slow growth, but increase tree mortality, changing the type of maintenance that needs to be performed. For more information on optimizing maintenance, refer to Chapter 2.

Tree growth regulators (TGRs) can significantly slow the growth rate of treated trees, potentially increasing the maintenance interval and reducing the overall cost of utility pruning. Integrating TGRs into utility pruning programs can also reduce the amount of biomass resulting from pruning and the risk posed by trees. For more information on TGRs, refer to Chapter 4.

Reclaiming Overgrown Facilities

If pruning is not performed at the optimum time, tree growth overtakes utility facilities. Such situations are most commonly caused by insufficient program funding. In a 1997 controlled study on three utilities in different parts of North America, Browning and Wiant determined that the cost of pruning operations in recovering facilities following deferral of maintenance is significantly higher than costs incurred in routine, cyclical maintenance (Table 3.1). Furthermore, the study found that a 20-percent reduction in funding would increase the maintenance interval (or cycle length) by 80 percent over 12 years, due to the added time required to prune overgrown trees. The researchers also pointed out that actual maintenance intervals would be further lengthened as available resources were directed to reactive work, and that additional costs would be incurred in service restoration and repair of damaged facilities (Browning and Wiant 1997).

The costs and consequences of deferred tree maintenance increase as trees grow to their natural shape and size, often becoming entangled with, and entirely hiding overhead utilities. This situation poses unnecessary risk to tree owners

Table 3.1 Projected impact of deferred maintenance on the average cost of pruning trees for line clearance. Values shown are U.S. dollars.

		Relative cost* to prune trees at a site that is				
Utility	Length of optimum clearance cycle	At optimum time**	1 year past optimum	2 years past optimum	3 years past optimum	4 years past optimum
A	5 Years	$1.00	$1.23	$1.43	$1.59	$1.69
B	5 Years	$1.00	$1.21	$1.39	$1.53	$1.64
C	6 Years	$1.00	$1.16	$1.30	$1.40	$1.47

*Excludes an adjustment for inflation.
**Optimum time is based on the industry standard of 10–15% maximum tree-to-conductor contact.
Source: Browning and Wiant 1997.

Dr. Alex Shigo

The late Dr. Alex Shigo had a profound influence on arboriculture. Many of his publications, including *A New Tree Biology and Dictionary* (1986) and *Modern Arboriculture* (1991), remain classics. His research resulted in groundbreaking advances in our understanding of trees, with principles such as compartmentalization of decay in trees (CODIT) and branch attachment. Out of those principles came arboricultural innovations like "natural target pruning" that were controversial when they were first advanced, but have become standard practice and remain so.

Shigo's influence extends to utility arboriculture as well. When his 1990 field guide, *Pruning Trees Near Electric Utility Lines* (also known as "the yellow book") was first published, rounding trees over was the standard practice at many utilities and was popular with customers. Shigo's field guide promoted natural target pruning, directing workers to "hit the targets," that is, to identify branches growing toward power lines, remove them entirely, and make final proper pruning cuts at a lead or lateral branch growing away from conductors. He declared that 90 percent of the time, removing three branches would provide 90 percent of necessary clearance—the "90-3-90" rule. This was important in the age of roundovers, when decades of repeated topping generated a few large limbs that became "sprout platforms" that dominated tree crowns. Back then, it was true that 90 percent of the time, removing three leads provided 90 percent of necessary clearance. However, it's less so now after a generation has applied natural target pruning to clear trees from power lines.

What's more, with his "yellow book," Shigo put the weight of his authority behind the need for line clearance. In so doing, he gave credibility to the profession of utility arboriculture—implicitly acknowledging that it is not only reasonable, but also necessary for utilities to prune trees clear of power lines if society expects safe, reliable electric power, and that there was a proper way to do it. Along

Alex Shigo.

the way, he subjected himself to criticism from those who resisted legitimizing utility arboriculture at all, and also from those who considered directionally-pruned trees to be an abomination. However, many utilities retained him to teach tree biology and directional pruning to their line-clearance crews (Tomosho 1998). Shigo's efforts expanded utility arboriculture as utilities increasingly realized that a professionally-run vegetation management program could lower costs and reduce outages; and the number of arborists working for utilities and their contractors increased markedly during the 1990s.

Utility arborists owe a debt of gratitude to Dr. Shigo. He brought the science of tree biology and modern arboricultural techniques to our profession, and many of us owe our careers to his efforts. Utilities, utility arborists and, most importantly, trees are better for it.

and passersby, reduces the reliability of utility service, and leads to longer restoration times following storms (United States Department of Energy 2012). When facilities become completely overgrown, extensive pruning and often complete tree removal is required (Figure 3.15). The appearance of trees and neighborhoods may change considerably as a result of these operations.

In addition to the higher costs of pruning and repair, utilities must overcome associated negative publicity and resulting customer complaints (S. Tankersley, personal communication). Further, liability for injury, fires, and property damage can be much larger where it is shown that inadequate vegetation management was a contributing factor (Tomasovic 2011). Finally, the cost of power outages for end users, while not borne directly by utilities, is a concern of utility customers and regulatory agencies.

To avoid the many problems inherent in recovering overgrown facilities, utility vegetation managers should strongly advocate for adequate funding to perform preventive maintenance at appropriate intervals, emphasizing key benefits, such as public and worker safety, reduced service interruptions, improved customer relations, and stabilized long-term costs (Sankowich 2013; Grayson n.d.).

PALM PRUNING

Members of the palm family *(Palmae)* are **monocots**, and are physiologically different from other trees. Upward growth in palms begins in the bud, which is located at the top of the stalk, just below the point where the **fronds** originate. If this bud is damaged or removed, then the remaining stem will die. For this reason, palms that are tall enough to affect utility facilities cannot be reduced in height, or directionally pruned (Figure 3.16). It is possible to remove fronds on one side of the plant, but the palm will quickly replace the missing fronds, so frequent pruning may be required if the palm is not removed. If the palm is retained in the landscape, it is important not to damage the trunk, as palms, unlike most other trees, have no means of closing such wounds (Hodel 2009).

Figure 3.15 Overgrown trees pose greater risk, require extensive pruning, and have been shown to drive up costs.

Figure 3.16 Palms cannot be reduced or directionally pruned, and thus pose unique risks to utility facilities.

Prune or Remove?

As a general rule of thumb, trees that cost as much or more to prune than to remove should be removed. This is more commonly the case when trees are relatively small. But there may be cases when pruning is preferred over removal, or vice versa, regardless of cost considerations. For example, trees in good condition that respond well to directional pruning and that pose little risk to utility facilities are likely adding value to adjacent properties and the community as a whole, and should not be removed. On the other hand, trees that pose unacceptable risk to facilities that cannot be mitigated by pruning, or trees that do not respond well to utility pruning and must be frequently and heavily pruned, are good candidates for removal (ANSI 2017b, 2017c). Tree owners may relocate young, incompatible trees as a practical alternative to removal, but only if the operation can be performed safely (Pacific Gas and Electric Company n.d.).

Removing trees that otherwise must be pruned many more times certainly saves future maintenance expense, even if the removal cost is considerably higher than pruning. However, it is important to consider whether the allocation of resources for tree removal will compromise other program objectives. Utilities sometimes allocate additional funds for targeted removal programs as part of an overall risk reduction or recovery program. In some cases, regulating agencies have allowed the use of capital funds for this purpose (Puget Sound Energy 2010).

These live oaks are providing canopy cover and many benefits to the community while posing little risk to utility facilities.

Utility arborists should be aware that even palms not immediately adjacent to facilities may pose a threat, especially during high winds, when flexible palm stems can sway, and fronds may break free and blow a considerable distance. The best solution for palms interfering with utility facilities is removal or relocation. Relocation of the palm may be undertaken by the property owner, but only if the operation can be safely accomplished without further impacting the facility (Pacific Gas and Electric Company n.d.).

REMOTE FORESTED ENVIRONMENTS

Utilities operating in remote areas often must maintain many miles of wooded utility corridors. In such high tree density areas, typical pruning practices described here may not be economical. In these settings, mechanical and chemical pruning methods are much faster and safer than manual methods, and are therefore recognized by industry

Support for Utility Vegetation Management

To generate more support for utility pruning and other vegetation management programs (Grayson n.d.):
- Communicate value to key stakeholders, internal and external.
- Focus on maximizing the value for the money spent (e.g., prioritizing risk-reduction efforts on the highest-priority areas).
- Design programs to address significant threats beyond the easement when necessary.
- Encourage employees to pursue professional credentials, through incentives and by paying or reimbursing fees.
- Adopt new and promising technologies to enhance efficiency.
- Adopt QA/QC procedures and pursue continuous improvement.
- Support industry efforts to improve practices through practical research and demonstration.

standards as acceptable, even though the quality of individual cuts may be compromised. (ANSI 2017b). However, the aesthetics of these operations should be taken into consideration, and their use limited in areas where complaints or negative reaction is likely (Brennan 2012). Corridors should be inspected regularly to ensure that pruned trees do not pose excessive risk. Required maintenance interval varies by treatment, species mix, and local growing conditions.

Mechanical Pruning

In many remote or rural areas, operators use equipment with large saws or cutter heads mounted on booms to prune limbs alongside right-of-way corridors. Saws are sometimes suspended from helicopters (Figure 3.17). Resulting brush may be dragged and stacked, or left in place, depending on local conditions and property owner concerns.

When undertaking mechanized pruning operations, an effort should be made to minimize injury to main stems by making cuts outside of the branch bark ridge and branch collar. Cut brush should not be left in areas where it could pose a fire risk, impede access, affect drainage, or where rising water could carry the material offsite (Fortin Consulting, Inc. 2014).

Figure 3.17 Mechanical pruning, shown here being performed using a helicopter, is acceptable in remote/rural areas, with the cooperation of landowners.

Chemical Pruning

Also known as chemical side trimming, the **chemical pruning** method utilizes certain herbicides selectively applied along the side of right-of-way corridors. Both foliage and buds are controlled, which results in a delay in re-sprouting. Treated branches eventually die and are naturally shed by the tree. The resulting browning of vegetation can be cause for public concern (Brennan 2012). Care must be taken to ensure that treatments are applied only to target branches. Excessive treatment is likely to kill the tree, and drip and drift may damage or kill off-target plants (Bayer AG 2018).

SUMMARY

While the pruning of vegetation away from utility facilities may at first appear to be a relatively simple endeavor, it is actually a complex process that is essential to ensure the safe and reliable delivery of utility services. Operating effectively in such a wide range of settings, and interacting with property owners, cities, land management agencies, and other stakeholders requires professional expertise in safety, arboriculture, communications, equipment operations, project management, and other specializations as required (Kligman 2012; Miller 2014).

Given that the benefits of utility pruning are so often taken for granted, it is not surprising that the efforts of utility arborists are not well understood by most members of the public. What is perhaps more surprising is the number of professionals in arboriculture, urban forestry, government, and even within the utility industry itself who are not fully aware of the value and importance of the service provided by utility arborists. To overcome this lack of awareness about the value of utility pruning, it is essential that practitioners from management to crew personnel understand the value of their work and perform as professionals.

CHAPTER 3 WORKBOOK

Fill in the Blank

1. An _____ is land over which the utility has a _____-___-_____ to maintain clearance.

2. The planned _____ _____ is the time between scheduled pruning operations.

3. When it has little or no likelihood of failure or impact to utility facilities, maintenance of the portion of a tree that is outside of the utility easement is the responsibility of the _____ _____.

4. Trees react to structural defects with _____ _____, which compensates for structural weakness.

5. Trees or branches that fall from outside utility easements have a _____ impact on service reliability than trees growing into and contacting power lines.

6. _____ uses reduction cuts to reduce branch length relative to other similar branches.

7. Trees that have been rounded over or topped in the past but later converted to directional pruning may have extensive _____ growth on decayed stubs.

8. As more of the tree crown is removed, subsequent shoot growth is likely to _____.

9. Maintenance that is performed before an occurrence such as a service interruption is termed _____, while maintenance that is performed in response to such an occurrence is termed _____.

Multiple Choice

1. All of the following are objectives of utility pruning *except*
   a. reducing risk
   b. maintaining tree structure
   c. providing clearance
   d. decreasing wind resistance

2. Utility easements may include only a portion of an individual tree.
   a. True
   b. False

3. Reducing or removing branches back to laterals or parent stems growing away from the facility is known as
   a. cycle pruning
   b. facility pruning
   c. directional pruning
   d. corrective pruning

4. Which of the following is true of palms?
   a. palms can be directionally pruned
   b. palms can be reduced in height
   c. palms can close over trunk wounds
   d. palms can quickly replace removed fronds

5. Which of the following is *not* true of tree growth regulators (TGRs)?
   a. TGRs are effective in reducing the rate of shoot growth
   b. TGRs have little or no deleterious effects
   c. TGRs result in leaves that are smaller and slightly darker than normal
   d. TGRs are most effective when applied to slow-growing trees

6. Chemical and mechanical pruning methods should be restricted to areas
   a. with high residential density and many trees
   b. where conventional methods are impractical, such as remote and rural locations
   c. where individual amenity trees must be reduced in height
   d. with strict regulations prohibiting arborists from climbing trees

7. Heavy sprout growth following utility pruning is never expected from healthy trees.
   a. True
   b. False

CHALLENGE QUESTIONS

1. Identify three things that could affect the extent of pruning specified in a given location and describe how work might be performed differently as a result.

2. Discuss how deferred maintenance can affect utility pruning programs.

3. What factors should be taken into account when determining the optimum pruning maintenance interval?

4. What are some ways to enhance the effectiveness of cycle maintenance?

5. Name five characteristics of branches that are more likely to fail.

4
Integrated Vegetation Management

OBJECTIVES

- Choose objectives based on the intended purpose of the site and available resources.
- Evaluate the site to assess field conditions.
- Compile and select from a broad array of treatment methods, including manual, mechanical, chemical, biological, and cultural.
- Identify engineering alternatives in vegetation management.
- Implement and monitor a vegetation management plan and efficacy of treatments.
- Describe key chemical properties of herbicides used in vegetation management.

KEY TERMS

action threshold
acute toxicity
allelopathy
basal application
border zone
chronic toxicity
closed chain of custody
compatible vegetation
comprehensive evaluation
corrosive
cover-type conversion
cover-type mapping
cultural control method
dose
hack and squirt
half-life

herbicide
incompatible vegetation
integrated pest management (IPM)
integrated vegetation management (IVM)
IVM methods
label
LC_{50}
LD_{50}
LiDAR
mechanical control methods
minimum vegetation clearance distance (MVCD)
mode of action
non-selective herbicides
peripheral zone

pipe zone
point sampling
poison
returnable, reusable (R/R) container
selective herbicides
setting objectives
stump application
supply container
tolerance level
toxicity
tree growth regulator (TGR)
tree risk assessment
wire zone
wire-border zone
workload assessment

INTRODUCTION

Integrated vegetation management (IVM) is a system of managing plant communities in which managers set objectives, identify compatible and incompatible vegetation, consider tolerance levels and action thresholds, and evaluate, select, and implement the most appropriate IVM method or blend of methods to achieve their established objectives. The choice of IVM methodology is based on their environmental impact and anticipated effectiveness, along with site characteristics, security, economics, current land use, and other factors (ANSI 2012). IVM best practices have the objective of establishing diverse, compatible plant communities. They are not a set of rigid prescriptions based upon established time periods, repeated unselective mowing or broadcast spraying across entire right-of-way widths.

IVM is a modification of the concept of **integrated pest management (IPM)**. The principal objective of integrated pest management is balanced use of control measures to maintain pest populations below tolerance levels. The "pest" in the case of IVM is vegetation incompatible with the intended use of the site. **Tolerance levels** can be quantified in terms of acceptable economic loss. Integrated pest management controls include mechanical, cultural, biological, chemical, and regulatory measures. Site characteristics, security, economics, current land use, environmental impact, the anticipated effectiveness of a technique, and other factors all help determine which approach is best suited to a particular situation (Miller 1993).

Nowak (2005) details a definition of IVM consistent with principles and practices of IPM. He characterizes IVM as a system designed to control incompatible vegetation to achieve specific management objectives and continual process improvement (Miller 2014). It is used to systematically choose, justify, selectively implement, and monitor different types of vegetation management treatments. Treatment selection is based on a control method's effectiveness, economic viability, environmental impact, and suitability for safety, along with site characteristics, security, socioeconomics, and other factors. IVM uses combinations of methods to promote sustainable plant communities that are **compatible** with the intended use of the site, and to control, discourage, or prevent establishment of **incompatible** plants that can create problems with safety, security, access, fire risk, utility service reliability, emergency restoration, visibility, line-of-sight requirements, regulatory compliance, or environmental integrity.

The key steps of IVM drawn from IPM include:

1. Gaining science-based understanding of incompatible vegetation and ecosystem dynamics
2. Setting management objectives that consider tolerance levels and action thresholds based on institutional requirements and broad stakeholder input
3. Selecting and applying treatments from a variety of options, including biological, chemical, manual, mechanical, and cultural methods to produce desired plant communities, with an emphasis on prevention through biological controls
4. Monitoring the system to determine the effectiveness and necessity of treatments in creating desired plant communities and achieving management objectives

The intent of IVM is sustainable management of a specific right-of-way based on balanced socioeconomic and environmental considerations.

The target of IVM is incompatible plants, which might be noxious weeds or invasive species; plants that pose potentially unacceptable economic, social, or environmental risk; or any vegetation that managers consider inappropriate for a given site. In a utility context, particular emphasis is placed on controlling plants that have the genetic capacity to interfere with or limit access to electrical or pipeline facilities.

Integrated vegetation management principles can be applied to all types of rights-of-way—including gas pipelines, highways, security corridors and canals, as well as to restoring ecosystems, controlling invasive weeds, and other actions (Miller 2014).

Successful integrated vegetation management requires planning and implementation. While planning and implementation in IVM have considerable overlap with program management, the distinction is that vegetation management focuses on the "how to do" while program management concentrates on "what to do."

In North America, the NERC *Transmission Vegetation Management Program Standard* (North American Electric Reliability Corporation 2008) requires utilities to document their transmission vegetation management maintenance strategies, procedures, processes, or specifications to prevent vegetation from intruding within the defined minimum clearance zone. This documentation can include written specifications, standard operating procedures, annual work plans, and the flexibility to adapt to changing conditions.

ANSI A300, Part 7 (2012) outlines integrated vegetation management. It is applicable to distribution as well as transmission projects and consists of six steps:

1. Set objectives
2. Evaluate the site
3. Define action thresholds
4. Evaluate and select control methods
5. Implement control methods
6. Monitor treatment and quality assurance

Decisions are required in setting objectives, defining action thresholds, and evaluating and selecting control methods. The process is cyclical (Figure 4.1), because managing dynamic systems is ongoing. Managers must have the flexibility to adjust their plans at each stage as new information becomes available and circumstances evolve.

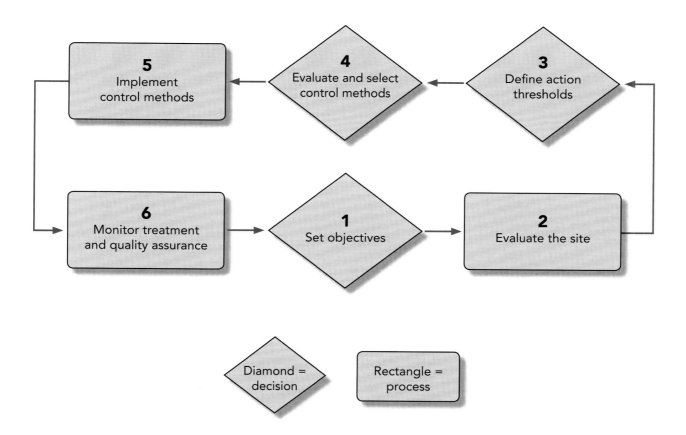

Figure 4.1 Integrated vegetation management flowchart. *(Courtesy Tree Care Industry Association, Inc. Visit TCIA.org for more information)*

SETTING OBJECTIVES

Setting objectives is the initial phase of the integrated vegetation management model. Objectives should be clearly defined and documented by the vegetation manager and be based on the intended purpose and use of the site. The availability of human, equipment, and financial resources will also determine objectives. Examples of common integrated vegetation management objectives include:

- Providing access
- Cost efficiency
- Cultural site protection
- Environmental protection
- Facility protection
- Reliability
- Safety
- Security

Often, the definitive objective is environmentally sound, cost-effective control of species that potentially conflict with or limit access to utility facilities, while promoting compatible, early successional, sustainable, plant communities. Vegetation managers should also consider concerns of landowners or natural resource authorities, provided those concerns do not conflict with primary IVM objectives.

Objectives for pipelines can involve safety, route identification, testing, encroachments, and maintenance and inspection, particularly aerial and ground patrol needed for leak detection. Route identification is especially important for underground facilities, which are often identified only by aboveground markers or valves, and measuring stations adjacent to the pipeline that are easily hidden by unmaintained, overgrown vegetation (Stedman and Brockbank 2012). A comparison of electric and pipeline rights-of-way concerns is presented in Table 4.1.

Objectives should be SMART (see Chapter 2): specific, measurable, achievable, relevant, and timely. The idea is to establish objectives that are precise and explain exactly what needs to be done, who needs to do it, and where it needs to be done. The objectives are measurable, so progress can be impartially evaluated. Unattainable or irrelevant goals are pointless, and timeliness requires deadlines to drive completion of the goal (DelPro 2007).

Wire-Border Zone Concept

A common objective of IVM is the **wire-border zone** concept, which is a management philosophy applied through cover-type conversion (covered later in this chapter). It was developed by W.C. Bramble and W.R. Byrnes in the mid-1980s out of research they started in 1952 on a transmission right-of-way in the Pennsylvania State Game Lands 33 Research and Demonstration project (Figures 4.2a and b) (Yahner and Hutnick 2004). It consists of a wire (wire security) border and peripheral zones.

The central consideration in setting zone boundaries and deciding the mature height of allowable species is to ensure that no plants will be allowed that have the capacity to grow into the tolerance zone at any time in their life. The **wire**

Table 4.1 Electric vs. pipeline right-of-way concerns.

Electric rights-of-way	Pipeline rights-of-way
Electric right-of-way identification is obvious with lines and tall structures	Pipeline right-of-way identification for underground facilities is less evident, because it is accomplished with markers, valves, and measuring stations that are easily obstructed by vegetation
Tree height under and to the side of lines, as well as distance to the side, affects safety and reliability	Trees block access and obstruct views
	Root intrusion (integrity of pipeline coating)

Source: Stedman and Brockbank 2012.

Figure 4.2 The Game Lands 33 Research and Demonstration project in 1953 (a) and in 2008 (b), located in Pennsylvania, U.S.

zone is the section of a utility transmission right-of-way under the wires and extending out both sides to a specified distance (Bramble et al. 1992). The standard way to establish the wire zone is by an established measure like 10 feet (3 m) or some other length to the side of the outside wires. The wire zone is managed to promote a low-growing plant community dominated by grasses, herbs, and small shrubs (e.g., under 3 feet [1 m] in height at maturity). The **border zone** is the remainder of the right-of-way (Figure 4.3), where small trees and tall shrubs (under 25 feet [8 m] in height at maturity) are established. The concept may be modified to accommodate side slope, where the border zone may need to be eliminated on the uphill and extended on the downhill side (Figure 4.4). When properly managed, diverse, tree-resistant plant communities, composed of species that will never interfere with utility facilities in their lifetimes, develop in wire and border zones.

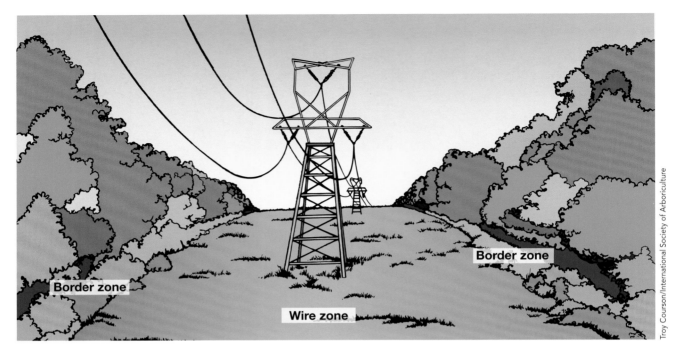

Figure 4.3 The wire-border zone includes compatible species throughout the right-of-way. The wire zone (directly under and just to the outside of wires) is managed to promote a low-growing plant community dominated by grasses, herbs, and small shrubs (perhaps under 3 feet [1 m] in height at maturity). The border zone (the remainder of the right-of-way) is where small trees and small shrubs comprising species that will never grow to interfere with tolerance levels (roughly 25 feet [8 m] at maturity) can be established.

The communities not only protect the utility facility and reduce long-term maintenance, but also enhance wildlife habitat, forest ecology, and aesthetic values. The area outside the right-of-way is the **peripheral zone**, where tall-growing species may be allowed, although they need to be monitored for risk.

While the wire-border zone is often a best practice, it is not necessarily universally suitable. For example, standard wire-border zone prescriptions may be unnecessary where lines are high off of the ground, such as across low valleys or canyons. One way to accommodate topography variances is to establish different plant communities based on wire height (Figure 4.5). For example, over canyon bottoms or other areas where conductors are an established height, such as 100 feet (30 m) or more above the ground, only a few trees, if any, are likely to be sufficiently tall to conflict with the lines. In such cases, trees that potentially interfere with the transmission lines can be removed selectively. In areas where the wire is lower, perhaps between 50 and 100 feet (15 and 30 m) from the ground, a border zone community can often be developed throughout the right-of-way. Where the line is less than 50 feet (15 m) off of the ground, managers could apply a full wire-border zone prescription. One environmental advantage of this type of modification is stream protection. Streams often course through the valleys and canyons across which lines are likely to be elevated. Leaving timber or border zone communities in these canyon bottoms helps provide riparian buffers and shelter this valuable habitat.

Strict adherence to wire-border zone methodology may also be inappropriate in some fire protection jurisdictions, where border zone establishment is discouraged out of concern it could provide ladder fuels to the adjacent forest. In these and other cases, management objectives could call for alternatives, such as a perennial meadow throughout the right-of-way. Meadows are legitimate, tree-resistant plant communities that can be established through integrated vegetation management. The point is that the wire-border zone is a useful concept where it meets management objectives. Whether or not it should be employed depends on the judgment of qualified vegetation managers.

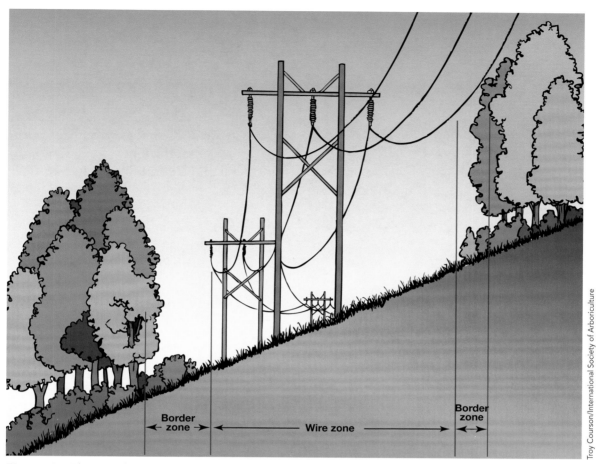

Figure 4.4 The wire-border zone can be modified to accommodate side slope.

Figure 4.5 The wire-border zone can be modified to accommodate topographical variances.

Pipe Zone-Border Zone

The wire-border zone concept can be modified to accommodate pipelines (Figure 4.6). Over the **pipe zone**, native forbs and grasses may be encouraged. Dense, low-growing, gas-sensitive, cover could also be introduced into the pipe zone if desired. The idea is to maintain visual sightlines to establish the pipeline corridor, facilitate access to the pipe with equipment, and prevent roots from encroaching on pipes (Porter and Cohen 2017). If prairie or other grasses are so tall that they interfere with testing or maintenance, a narrow path directly over the pipe can be mowed without disturbing the remainder of the right-of-way. This would result in periodic strip-mowing as needed, with low economic and environmental costs and greater benefits for certain wildlife species. Border zone vegetation can be managed on the edges of the pipeline right-of-way, where it will not interfere with maintenance or pipe integrity (Johnstone and Jaggie 2012; Stedman and Brockbank 2012).

Define the Scale of the Plan

Planning can range from establishing systematic strategies on one hand, to setting detailed, tactical, operational requirements for individual projects on the other. Data can be applied to establishing or modifying objectives, setting budgets or determining human, material, and equipment resources requirements (Miller 2014).

Urban forestry greenspace inventory techniques described by Miller and colleagues (2015) can be applied to utility arboriculture. Greenspace inventories can either be comprehensive or sample-based, depending on management needs, variability of the plant community or available

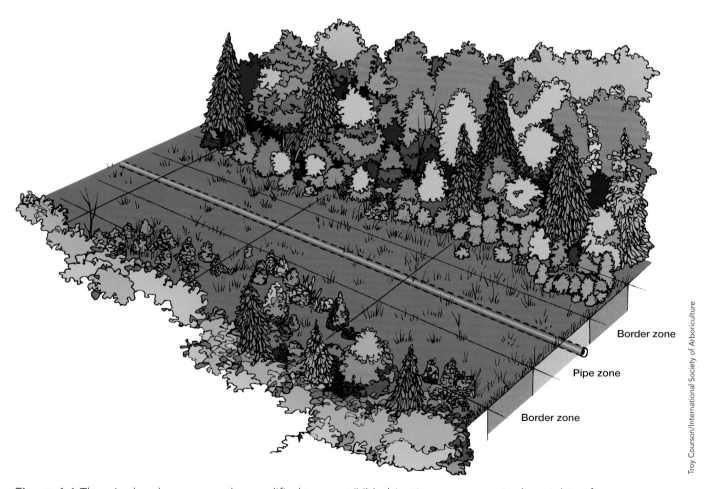

Figure 4.6 The wire-border zone can be modified to meet IVM objectives on many pipeline rights-of-way.

funds. **Cover-type mapping** is a potentially useful technique, which uses aerial photographs or satellite imagery followed by ground checks to determine the nature of plant communities on a site. Internet sites providing satellite photos can be applied to cover-type mapping, particularly as an overlay with plan and profiles on geographic information system (GIS) platforms (Figure 4.7). The National Land Cover Database, which offers satellite imagery data on cover types and land cover transformation of the U.S. over time (Homer et al. 2015), is an example of such a site. Ground checks are useful to confirm information on landforms, vegetation, soils, and aquatic or riparian delineation. Ground inventories can also provide specific data to verify general information found on the maps (e.g., species and descriptions of ecotypes and understory), or be used in the absence of recent photos (Miller et al. 2015).

Comprehensive Evaluations

Comprehensive evaluations account for all vegetation that could potentially affect management objectives. Program level comprehensive evaluations can be made of all target vegetation on a system. Those evaluations at a project level focus on vegetation relevant to a specific job. Either way, comprehensive evaluations provide the advantage of supplying a complete set of data upon which to base management decisions. Comprehensive surveys are uncommon for utilities, though, because of the large numbers of trees.

Increasingly, utilities are using light detection and ranging (LiDAR) to evaluate workloads on rights-of-way. **LiDAR** transmits laser pulses toward a target, records the time it takes to return to sensors, and translates the results into a virtual image of the right-of-way (Figure 4.8) (NOAA n.d. [b]). The technology is precise and can document the distance between trees and conductors, as well as calculate how far that distance will fluctuate under environmental conditions and specified design parameters. LiDAR can acquire data by air, ground or both. It can identify trees that are within striking or arcing distance of lines should they [the trees] fall. LiDAR can also be combined with global information systems to locate structures, as well as line location, right-of-way information, topography, and line movement calculations on maps. The package produces drawings in a format that can be

Figure 4.7 Tree information shown on a geographic information system platform.

Figure 4.8 LiDAR computer image.

taken into the field for use by tree crews (Hooper 2003). While LiDAR cannot differentiate among species, technology is being developed to identify different types of trees (i.e., deciduous or conifer) or distinguish between living and dead trees. However, it does not replace ground survey verification, which is necessary to detect tree defects or soil conditions.

EVALUATING THE SITE

Site evaluations are used to assess field conditions for planning. The site in utility terms is often a distribution or transmission right-of-way, but it can also be substation, generation, operational, fee-owned, or other property. Site evaluations can identify a variety of factors such as potential safety issues, applicable regulations, workload, line voltage or pipeline capacity and criticality, funding, labor, and equipment resource availability, height of the wire from the ground, right-of-way width, land ownership and use, fire risk, vulnerable or protected areas, presence of species of concern, water resources, archeological or cultural sites, topography, soils, and other matters.

Evaluations provide information on site characteristics that exist at the time an assessment is conducted. On dynamic systems, such as those associated with IVM, information can quickly become outdated, meaning that regularly scheduled updates, condition assessments, and inspections will be required. Schedules should be based on anticipated vegetation growth, line design and construction, predominate species, environmental factors, political considerations, budgetary parameters, and operational issues.

Careful preparation is needed to ensure that valuable time and resources are directed toward obtaining useful information and not wasted on collecting unnecessary details (Miller et al. 2015).

Examples of factors to identify in site evaluations include:

- Access routes
- Anticipated control method(s)
- Applicable regulations
- Archeological and cultural issues
- Fire risk
- Funding availability
- Height of the wire from the ground
- Labor and equipment resource status
- Land ownership and use
- Line construction
- Line location
- Line voltage and criticality
- Pipeline capacities
- Presence of species of concern
- Right-of-way width

Environmental Protection

Species of concern. Utility vegetation managers need to adhere to applicable guidelines and regulations. Vegetation management should not disturb or harm species of concern (i.e., rare, threatened, endangered, or otherwise protected species). Often, simple adjustments can be made to protect sensitive species without compromising desired outcomes.

Wetlands. Wetlands should be worked using suitable control methods. If herbicides are to be applied, only those labeled for use over water may be used in wetlands.

Stream protection. To protect streams, incompatible vegetation may need to be selectively removed, or treated with appropriate herbicide to gradually establish a compatible riparian plant community. In some cases, pruning trees may allow them to be retained, if appropriate. Equipment may only use existing or designated stream crossings.

Stream crossings of right-of-way corridors, surface water supply reservoirs, and drinking water wells and springs need to be protected by buffers. Buffers should retain as much compatible vegetation as possible. If herbicides are needed within the buffer, only those appropriate for the site should be applied. Machine work should be avoided in buffers as equipment may cause erosion or leak or spill petroleum products, causing pollution. Utility vegetation managers working along with competent authorities should determine appropriate distances for particular buffers.

Archeological or cultural sites. Vegetation management activities should not disturb known archaeological or cultural sites. When necessary, archeological sites should be located and marked, and a plan established to adequately protect them during work. Field data inventories of known sites should be kept on file, preferably accessible through a GIS system. Practices that won't damage the sites, such as manual cutting and backpack or aerial herbicide applications, should be considered for use at these locations.

- Safety concerns
- Soil types
- Topography
- Vulnerable or protected areas
- Water resources
- Workload

Workload Evaluations

Workload evaluations are inventories or surveys of vegetation that could have a bearing on management objectives. Understanding workload not only means quantifying what needs to be done, but also understanding the characteristics of the target vegetation, such as tree size, growth within work thresholds, species, and other factors.

Depending on objectives and available resources, utilities can either conduct comprehensive or partial evaluations. Partial evaluations or surveys are more common in vegetation management than they are in urban and community forestry. **Workload assessments** can collect data on an array of vegetation characteristics, such as location, height, density, species, size, and condition, **tree risk assessment**, and clearance from conductors. Evaluations should be conducted considering voltage, conductor sag from ambient temperatures and loading, the potential influence of wind on line sway, pipeline capacity, and other factors.

Urban forestry inventory principles can be adapted to utility arboriculture. Miller et al. (2015) observe that urban forestry inventory needs vary with the size of the community, level of service desired, and potential vegetation problems. Likewise, the size of the utility or management area, commitment to service, and potential vegetation problems will influence workload assessments for utilities. Some programs will function well with samples covering only a few parameters, while others will require sophisticated, comprehensive inventory systems.

In a classic article on designing urban forest inventories, Zeisemer (1977) recommended a systematic, decision-making process involving the answering of the following seven questions, which can be applied to IVM:

1. What kind of shape is the system in?
2. How is work being assigned?
3. What are your objectives?
4. Is the program serving the customers well?
5. How much work is being done systematically versus responding to requests?
6. Is the program funded appropriately?
7. Which information is important and which is not?

Zeisemer points out that it is important to know which tree-related information is needed at various levels in the organization, when it is needed, and for what purpose.

For utility application, managers can apply Zeisemer's approach and draw data on a variety of vegetation characteristics, such as:

- Clearance from conductors
- Condition
- Density
- Diameter
- Growth rates
- Height
- Location
- Species
- Tree risk assessment status
- Unacceptable overhang

Assessments should be conducted considering circuit voltage, conductor sag resulting from high temperatures or heavy electrical loads, and the potential influence of wind on line sway.

Tree Risk Assessment

Many utilities manage hundreds of trees for each mile of right-of-way, which, across entire systems, can add up to millions of trees. With these populations, it is unreasonable to expect utilities to precisely identify and mitigate the risk posed by every individual tree. To further complicate matters, utilities may be hindered by property owner or land management agency opposition. Therefore, the only plausible course of action is for utilities to manage, rather than eliminate, risk from trees (Utility Arborist Association 2009).

To systematically manage tree risk to facilities, utilities should use a defined methodology for inspection and documentation, such as that described in ANSI A300, Part 9 (ANSI 2017c) and ISA's Tree Risk Assessment Qualification (TRAQ) program (Figure 4.9). A Level 1 or limited visual assessment of trees along utility corridors can be incorporated into most utility inspection and maintenance programs. Level 1 assessments review large populations of trees from a specified perspective to identify trees with obvious defects, which are often those with a relatively high likelihood of failure. Assessments may be conducted by walk-bys, drive-bys (a.k.a. windshield), or aerial patrols (Smiley et al. 2017).

Level 1 assessments are by definition limited and cannot identify all significant tree defects. However, Level 1 inspections will identify many individual defects, as well as other conditions of concern, allowing managers to better focus mitigation efforts (Smiley et al. 2017). If an initial Level 1 assessment identifies a need for greater scrutiny, a Level 2 or basic assessment may be necessary. A Level 2 assessment is a detailed, 360-degree ground-based visual inspection of the aboveground portion of a tree and its surrounding site, often using simple tools (Table 4.2). Level 2 assessments are also limited in that they may not always detect certain internal, belowground, and upper crown defects (Smiley et al. 2017). Level 3, or advanced assessments, while occasionally required to further evaluate defects not detectible with level 2 assessments, are impractical in most utility settings (Utility Arborist Association 2009).

Trees that pose an unacceptable level of risk require a mitigation plan. Each utility should have

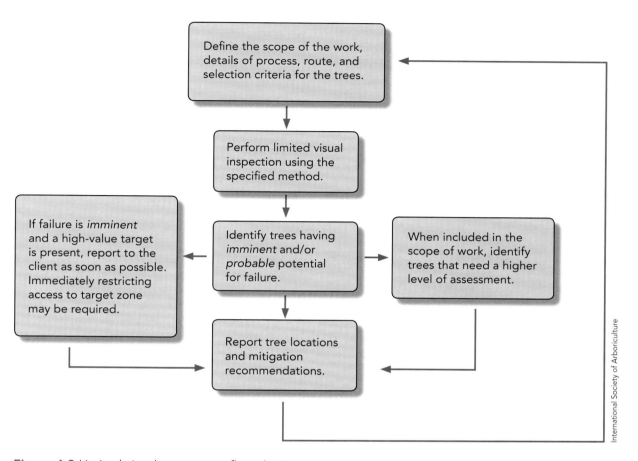

Figure 4.9 Limited visual assessment flow chart.

Table 4.2 Equipment that may be used in basic assessments.	
Equipment	**Use**
Binoculars	Examine the upper portions of a tree's crown for cavities, nesting holes, cracks, weak branch unions, and other conditions and tree responses
Magnifying glass	Help identify fungal fruiting bodies or pests that may affect the overall health of the tree
Compass	Determine the orientation of the tree, site, and defects or conditions on the tree being assessed
Camera	Document conditions or use "zoom in" feature to better observe conditions in the canopy of a tree
Mallet (made of wood, rubber, leather, or resin)	Sound the trunk for indications of hollows or dead bark
Probe	Examine the extent of cavities
Trowel or shovel	Conduct minor excavations of roots or the root collar (extensive root or root collar examinations are outside the scope of basic assessments)

Source: Smiley, Matheny, and Lilly 2017.

a plan and procedure for assessing and addressing high-risk trees. The plan should assign responsibility for its execution (Utility Arborist Association 2009). When trees that pose an imminent threat to subject transmission facilities are identified, FAC-003-4 (North American Electric Reliability Corporation 2016) requires transmission owners to notify the appropriate switching authority that vegetation is likely to cause an outage at any moment.

Utility arborists interested in more detailed tree risk assessment information are directed to *Utility Best Management Practices Tree Risk Assessment and Abatement for Fire-prone States and Provinces in the Western Region of North America* (Utility Arborist Association 2009), ISA's *Best Management Practices: Tree Risk Assessment* (Smiley et al. 2017), and the *Tree Risk Assessment Manual* (Dunster et al. 2017).

Partial Tree Evaluations

Expansive geographic territory and the large number of trees on many systems can make comprehensive workload evaluations prohibitively expensive and time consuming. For those programs, partial evaluations or surveys might be an economical and practical way to gather accurate management information. While partial evaluations are inappropriate for tree risk assessment and mitigation, they can be used to project the total amount of work from a representative population. They are cost-effective, and have a proven track record for reasonable accuracy.

Ecologists and foresters have developed a variety of time-tested sampling methodologies that can be applied to utility arboriculture. Completely randomized sampling provides the best opportunity for a representative population (Smith 1980). It gives each unit an equal probability of being selected, thereby eliminating the chance of bias on the part of the investigator. Partial evaluations often require statistical analysis and can involve techniques such as point sampling, timber cruising, or mechanical sampling, among others.

Quadrat sampling is a type of partial evaluation using specified sample plots. The size and number of sample plots is determined by statistical requirements, and will vary depending on the objectives of the study and the type of plant community (Curtis 1971). Management areas (whether a system or project) are often divided into equal-sized units and a statistically representative sample randomly selected for evaluation. Every plant or plant community of interest within each selected area is inventoried, with collected data extrapolated to forecast total workload.

Point sampling is frequently used in traditional forestry to determine standing timber volume. It involves a random sample that allows every part of the forest an equal and independent chance to be recorded. The technique can be adapted for utility application. In a utility context, this technique is not spatially random, but designed around a utility infrastructure target. One approach is to divide a line section into 100 even parts, draw 10 at random, and inventory the selected areas for target vegetation. Mechanical sampling is another alternative, where sample plots are located along a designated line and established at predetermined intervals for a given percentage sample (e.g., a 100-meter length along an entire right-of-way), and vegetation within each plot documented (Miller et al. 2015).

DEFINING ACTION THRESHOLDS

Tolerance levels are maximum incompatible plant pressures (species, density, height, location, or condition) allowable before unacceptable consequences develop. **Action thresholds** are vegetation pressures where vegetation management treatments should occur to prevent conditions from reaching tolerance levels (ANSI 2012). Set time periods, such as cycle lengths, can make unsuitable action thresholds, particularly on transmission facilities, because variability in growth rates, conditions, the importance of the line in terms of the number or types of customers served, land development, and other factors will often cause work-timing requirements to fluctuate (Miller 2014). Tolerance levels and action thresholds will vary among utilities, and within a utility among projects. For example, encroachment is far less tolerable for transmission than distribution facilities, so it is reasonable to set tighter tolerance levels and action thresholds on transmission than distribution lines. The variability demands judgment, so tolerance levels and action thresholds must be established by qualified vegetation managers.

To establish tolerance levels and action thresholds, practitioners must at least consider the potential growth of vegetation, the combined movement of vegetation and conductors in high wind, and the sag of power lines due to elevated temperatures or icing.

Minimum Clearances

Minimum clearance requirements may be established by regulatory oversight or individual utilities to achieve management objectives. However, achieving mandated minimum vegetation clearance distances (such as the **minimum vegetation clearance distance [MVCD]** in FAC-003-4 [North American Electric Reliability Corporation 2016]), while technically in compliance with regulations, is not a suitable tolerance level. Establishing minimum clearances as an action threshold allows establishment of incompatible trees on the right-of-way, which would require periodic topping or severe pruning. In addition to creating unacceptable ongoing risk to facilities, it can unnecessarily place tree maintenance workers at risk.

Managers should bear in mind that clearances are just one objective out of many. IVM requires a broader, more preventative approach than simply maintaining minimum clearances. The objective of most IVM programs includes preventing incompatible vegetation from getting established in rights-of-way altogether. Trees that have grown to the point where air-gap spark-over or an interruption of service is likely at any moment indicate a breakdown of the IVM program. Therefore, tolerance levels and action thresholds should be used to manage incompatible vegetation long before it has the potential to violate minimum clearance requirements.

EVALUATING AND SELECTING CONTROL METHODS

IVM methods are the processes through which managers achieve objectives. The most appropriate control methods are those that are best suited to a particular site. Many cases call for a combination or blend of methods. Managers have a variety of controls from which to choose—including manual, mechanical, chemical (herbicide and tree growth regulators), biological, and cultural options. The ultimate objective is to maintain a compatible plant community with available tools, emphasizing biological and cultural control.

Manual Methods

Manual methods are performed by workers using hand-carried tools, such as chain saws, handsaws, pruning shears, and other devices (Figure 4.10). The advantage of manual techniques is that they are selective and can be used where others may not be appropriate, including urban or developed areas, environmentally sensitive locations (such as wetlands or places inhabited by protected species), near archeological sites, on steep terrain, and other situations. However, manual techniques can be inefficient and expensive compared to other methods. Contrary to common belief, manual techniques are not without potentially negative environmental effects. For example, if chain saws are used, they can cause soil and water contamination from gasoline spillage, lubricant spray from

Figure 4.10 Manual control methods are performed by workers using hand-carried tools, such as chain saws.

Tree Pruning and Removal

Electric distribution lines are often maintained with pruning as a part of an overall IVM strategy. Pruning for clearance of trees within pipeline and electric transmission rights-of-way can be inconsistent with IVM management objectives. However, it may be necessary in cases involving legal restrictions or on trees adjacent to the right-of-way. When pruning is necessary, it should be conducted according to the most current version of the ANSI A300 pruning standard (ANSI 2017b) and ISA's *Best Management Practices: Utility Pruning of Trees* (Kempter 2004). Structurally unsound or dead trees located off the right-of-way in remote areas may be left for wildlife by reducing them in height so they will no longer strike the electric facility if they should fail.

chains, noise and air pollution, and increased risk to worker safety.

Pruning is a common manual technique (and cultural method), particularly on distribution lines. However, in many cases, pruning does not meet management objectives for transmission lines. Refer to Chapter 3 for detailed information on utility pruning.

An aerial lift is a truck, logging skidder, or other vehicle mounted with a hydraulic boom-supported "bucket" or semi-enclosed working platform, often used to elevate workers. They can provide tree workers with access to trees to apply manual control methods (Figure 4.11).

Mechanical Control Methods

Mechanical control methods utilize machines that cannot be hand-held. The advantage of mechanical control is potential efficiency. It can also be cost-effective, at least in the short term, particularly for clearing dense vegetation during initial establishment, or reclaiming neglected or overgrown rights-of-way. On the other hand, mechanical control methods can be non-selective and disturb wildlife or sensitive sites, like wetlands, archeologically or culturally rich locations, or developed areas. Heavy equipment can be unstable and its use risky on steep terrain. Mechanical cutting also has inherent safety concerns with sharp cutting blades and rapid discharge of severed wood. Furthermore, machines can create ruts in wet areas and leave behind petroleum products from leaks or spills, particularly if they are poorly maintained.

Routine use of mechanical control by itself is probably better characterized as vegetation *maintenance* than vegetation *management*, since it seldom converts incompatible plant communities to compatible cover types. The problem is that mowed vegetation still retains its roots, which proliferate resurgent shoot growth that increases stem density in many species (Figure 4.12). So, repeated cutting is only effective in the short term, and is often counterproductive and not cost-effective unless combined with herbicide application.

There are many machines that can be used for integrated vegetation management. Types include brush cutters, shears, aerial lifts, pruning machines, and other equipment.

Brush-cutting equipment not only effectively removes and grinds brush, but can also fell small trees. It is often designed with rotary cutting heads or flail cutters. Rotary cutting heads have one or more blades that rotate and cut or shred vegetation, much like stump cutter heads. Flail mowers knock down and grate vegetation with rotating drums fitted with metal teeth, chains or anvil cutting heads (Figure 4.13). The resulting mulching and scattering helps accelerate debris decomposition, reduce fuel loads, and minimize fire risk. Appropriate timing and frequency can affect plant community development.

A feller-buncher is a whole-tree removal device mounted on an excavator or other heavy equipment. It can fell, lift, and stack trees (Figure 4.14).

Figure 4.11 Lifts mounted on all-terrain vehicles can access off-road locations.

Figure 4.12 Mowed vegetation proliferates resurgent shoot growth that can increase stem density in many species.

Figure 4.13 Brush-cutting equipment removes and grinds brush and fells small trees.

Figure 4.14 A feller-buncher can fell, lift, and stack trees.

Cut stumps of some trees can re-sprout with fast-growing shoots from the root collar or cambium layer, in the same manner as those resulting from mowing. Herbicide treatment to stumps or to subsequent sprouts is important to gain full control.

A pruning machine is a vehicle mounted with a telescoping boom (e.g., 75 feet [25 m]) fitted with a circular saw head (Figure 4.15). Mechanical cutting can also be done with an array of blades slung under a helicopter. These devices can prune trees quickly and efficiently; however, it can be difficult to hit natural targets with mechanized pruning equipment. Wounds that result are inappropriate for landscape or high-value trees. Consequently, ANSI A300, Part 1 (2017b) recommends limiting mechanical pruning equipment to rural or remote areas.

Chemical Control Methods

Chemical control methods include tree growth regulators and herbicides, and are essential for many vegetation management situations, particularly when combined with other methods.

Tree Growth Regulator Application Methods

Tree growth regulators (TGRs) are chemical products that slow the growth rate of plants, including trees. When properly applied, they are effective in reducing the rate of shoot growth with no deleterious effects (Figure 4.16). In addition, they have been shown to benefit tree health (Cheney 2005). Due to reduced shoot growth, the leaves of treated trees can be smaller and slightly darker than normal (Redding et al. 1994).

Tree growth regulators are applied in three ways: soil drench, soil injection, and trunk injection. Doses are calibrated according to a number of factors, including tree species, vitality, overall size, trunk diameter, and crown size. Therefore, it is important that employees are able to accurately measure tree size, recognize different species, and

Figure 4.15 Mechanized pruning machines are booms fitted with a circular saw blade.

properly use the application equipment (Cheney 2005; T. Prosser, personal communication).

- Soil drench is a technique in which a shallow depression (a couple of centimeters or about an inch deep) is dug around the base of a tree, then filled with tree growth regulators (Figure 4.17).
- Soil injection is performed by inserting a probe and injecting growth regulators into the soil at set distances apart around the base of the trunk under the drip line (Figure 4.18).
- Trunk injection is accomplished with a small diameter injector, which delivers tree growth regulators directly into the xylem (Figure 4.19).

By reducing tree growth rates and the size of trees adjacent to utility facilities, TGRs potentially leverage additional benefits that lead to further cost savings, including (Redding et al. 1994; Bai et al. 2004; T. Prosser, personal communication):

- Reduced variability in growth rates of tree populations

Figure 4.16 Tree growth regulators are effective in slowing the growth rate of trees. Branches treated with TGR (left); branches not treated with TGR (right).

Figure 4.17 Soil drench TGR application.

- Longer intervals between pruning operations
- Reduced time required for future pruning
- Healthier tree population with less likelihood of failure and impact to surroundings
- Less damage should trees fail
- Less debris to chip and haul
- Reduced customer concerns about the amount and effect of utility pruning

Herbicide Application Methods

Herbicide application methods are categorized by the quantity of herbicide used, the character of the target, vegetation density, and site parameters. Dyes can be added to the herbicide mix to mark areas that have been treated (Figure 4.20). Treatment includes individual stem, broadcast, and aerial methods.

Figure 4.18 Soil injection TGR application.

Figure 4.19 Trunk injection TGR application.

Figure 4.20 Dyes can be added to the herbicide mix to mark areas that have been treated, in this case, during stump application.

Individual Stem Treatments

Individual stem treatments are selective applications. They include stump, basal, injection, frill ("hack and squirt"), selective foliar, and side-pruning applications (Table 4.3). Because they are applied selectively, proper individual stem applications work well to avoid damage to sensitive or off-target plants. However, they are impractical against broad areas or for sites dominated by undesirable species.

Table 4.3 Herbicide treatment methods.

Individual Stem	Broadcast	Aerial
Basal	Bare ground	Fixed wing
Frill	Cut stubble	Rotary wing
Injection	High-volume foliar	
High-volume selective foliar	Low-volume foliar	
Low-volume selective foliar		
Side-pruning		
Stump		

- **Stump applications** are a common individual stem treatment, where herbicides are applied to the stump surface around the cambium and top side of the bark. Water-based formulations require immediate treatment, while oil-based herbicides penetrate sufficiently well to be applied hours, days, or even weeks after cutting.

- Injections are done by inserting herbicide into trees. While frill (commonly called **hack and squirt**) treatments involve applying herbicide into cuts in the trunk, injections or frill treatments are especially useful against large, incompatible trees that can be left standing for wildlife.

- **Basal applications** are often made to the base of the stems and root collar with a herbicide in an oil carrier (Figure 4.21). The oil penetrates the bark and carries herbicide into the plant. Although basal applications can be made year-round, dormant treatment is often best on deciduous plants, when they do not have foliage that can obstruct access to individual stems.

- Selective foliar applications involve spraying leaves and shoots of specific target plants, and can be either low- or high-volume. For low volume applications, comparatively high concentrations of the herbicide's active ingredient are made in less water than would be used in high-volume treatments. Selective foliar applications can also be made with thin paraffin inversions. Foliar applications are only made during the active growing season, normally in late spring to early autumn.

- Chemical side-pruning application is a technique where specific, non-translocatable herbicides are applied to control branches growing toward utility facilities. Treating large branches could damage trees in the same way as removing them through pruning, or due to overapplication in doses sufficiently high to damage or kill trees. It is, however, an effective treatment against smaller branches that have sprouted due to mechanical or manual pruning.

Broadcast Treatments

Broadcast applications are non-selective because they control all plants that are sensitive to the particular herbicide being used. Herbicide choice can provide an additional degree of selectivity with selective chemicals. Even then, broadcast treatments do not differentiate between compatible and incompatible plants that are sensitive to the herbicide. Broadcasting is particularly useful against large infestations of incompatible vegetation (including invasive species) in rights-of-way or along access roads.

Broadcast techniques include high-volume foliar, low-volume foliar, cut stubble, and bare ground applications.

- High-volume foliar application is similar to high-volume selective foliar application. The difference is that broadcast high-volume foliar treatments target a broad area of incompatible

Figure 4.21 Basal applications are made to the base of stems and root collars with a herbicide in an oil carrier.

species, rather than individual plants or pockets of plants.

- Low-volume foliar broadcasts a calibrated rate of 5 to 25 gallons an acre (approximately 45 to 230 liters per hectare) of herbicide solution in water or paraffin-oil thin inversion mix.

- Cut-stubble applications are made over areas that have just been mowed, using herbicides to prevent incompatible species from sprouting back from their roots. This application is usually restricted to the wire zone of an electrical transmission corridor in forested settings (see the *Wire-Border Zone Concept* section). Limiting spraying to the wire zone provides a buffer that reduces the risk of root uptake of herbicides by non-target trees adjacent to the right-of-way.

- Bare-ground treatments are used for clearing all plant material in a prescribed area, such as in substations or around pole bases (Figure 4.22). Bare-ground applications are usually granular or liquid applications following mechanical removal of vegetation, or used as a pre-emergent in early spring for maintaining graveled areas (e.g., substations).

Aerial Treatments

Aerial treatments are made by helicopter (rotary wing) or small airplane (fixed wing). Treatments via rotary wing aircraft are more accurate because helicopters can fly more slowly and are more maneuverable than airplanes. Airplanes are less expensive to operate than helicopters, but their comparative lack of maneuverability often makes them unsuitable for treating narrow utility rights-of-way, so they are rarely used. Unmanned aerial vehicle (drone) technologies are being developed for use in chemical application in small areas (Krause 2018). Aerial applications can be useful in remote or difficult-to-access sites. Where incompatible vegetation dominates a right-of-way, they can also be cost-effective and fast, especially if large areas need to be treated. Aerial control methods are non-selective, although they can provide a level of selectivity with proper herbicides. The primary disadvantage of aerial application is that it carries the threat of off-target drift. As a result, it must be performed under low-wind conditions with slightly toxic or relatively nontoxic herbicides, using valve boom sprayers or nozzles with controlled droplet sizes.

Biological Control Methods

Biological control is management of vegetation by establishing and conserving compatible, stable plant communities using natural competition, animals, insects, or pathogens. Biological control should be the preferred control method wherever possible. Some plants, including certain grasses, release chemicals that suppress other plant species growing around them. Known as **allelopathy**, this characteristic can serve as a type of biological control against incompatible species. Allelochemicals can inhibit seed germination, disrupt plant

Figure 4.22 Bare-ground treatments are used for clearing all plant material in a prescribed area, such as around pole bases for fire prevention.

physiology, damage root hairs and hinder mycorrhizae (Chick 2010). Promoting wildlife populations is another form of biological control because birds, rodents, and other animals often prefer to eat seeds or shoots of undesirable plants, which encourages compatible plant communities (Yanner and Hutnik 2004).

A chemically-facilitated biological control known as **cover-type conversion** provides a competitive advantage to short-growing, early successional plants, allowing them to thrive and successfully compete against unwanted tree species for sunlight, essential elements, and water. The early successional plant community is relatively stable and tree-resistant. It reduces the necessary amount of work to manage, including herbicide application, with each successive treatment. While it is a type of biological control, cover-type conversion may require the use of one or more other control methods, such as manual, mechanical, or chemical, depending on circumstances. The wire-border zone concept leverages cover-type conversion (see the *Wire-Border Zone Concept* section).

Tree-resistant communities are often created in two stages. The first involves non-selectively clearing the right-of-way of undesirable trees using the best applicable control method or methods. The second develops a tree-resistant plant community using selective techniques, often including herbicide applications and releasing the seed bank of native, compatible species lying dormant on site. In the long run, this type of biological control is the most desirable method, at least where it can be done effectively (Miller 2014).

Cultural Control Methods

Cultural control methods modify habitat to discourage incompatible vegetation and establish and manage compatible plant communities. It often requires more intervention to maintain than biological control. Examples of cultural control include pruning, seeding, or planting lawns or low-growing crops, establishing pastures, prairies, and compatible landscapes, as well as other managed areas. Fertilization and irrigation are types of techniques that may be used to help establish low-growing, compatible plant communities (Miller 2014). Vera-Art (2017) described a cultural control involving hydroseeding that can be applied after new construction, fire, or by disking or tilling prior to organic seeding. Plant communities established through hydroseeding can develop into biological controls. Chick (2010) has suggested actively planting species highly resistant to tree invasion as a cultural transition to biological control.

Engineering Solutions

While not vegetation control methods, engineering solutions can provide relief from vegetation-power line conflicts. Solutions may include relocating lines or putting them underground. A disadvantage of engineering solutions is that they are often unaffordable for adjacent property owners. They can also have detrimental environmental impacts if inappropriately applied (Goodfellow 1995). These shortcomings should be taken into account when determining applicability.

Goodfellow (1995) describes a variety of engineering solutions, such as installing underground lines, avoiding trees by constructing lines across a street and back again, and compact construction using covered overhead primary (Figure 4.23) and Hendrix spacers (Figure 4.24). Engineered solutions can also involve modifying infrastructure through strategies like changing overcurrent protection, moving or raising poles, installing alley arms, or replacing compact construction with crossarms (Table 4.4).

Underground Construction

On its surface, underground construction seems to be an obvious solution to tree–power line conflicts, and it often is. However, it is not a panacea; underground construction carries problems of its own.

Goodfellow (1995) outlines the advantages and disadvantages of underground construction (Table 4.4). One advantage is that many customers prefer underground construction because it is out of sight and can be subject to fewer tree-related outages than overhead configuration. On the other hand, underground electrical lines compete for space with other underground utilities, such as gas and water. It can be at least five times more

Figure 4.23 Compact construction with a covered primary.

Figure 4.24 An aerial Hendrix spacer system.

expensive to construct for distribution and much more for transmission facilities. Often utilities can perform vegetation management on trees for many decades for the cost of installing underground facilities. If the objective of putting lines underground is to protect trees, perhaps that outcome would be better served by taking the money that would be needed to reconfigure the electrical grid, and devoting it directly to urban and community forestry.

Moreover, underground lines may only provide marginal improvement in electrical reliability, where a fully functional overhead system is already in place. Underground outages are often lengthier than overhead outages. While the durability of underground cable has recently increased, its lifetime is still limited. Not only can replacement be expensive, but it can also disturb landscapes that have been in place for decades.

Installing underground utility facilities can harm trees. Tree roots in many soil types (particularly in the compacted soil of urban areas) are near the surface, with roots most often in the top 3 feet (1 m) of soil. They can spread out a distance of three times the height of the tree or more (Harris et al. 2004). Underground utility facilities are often buried 4 feet (1.3 m) below the ground. Trenching to install underground facilities can sever root systems and damage trees if the trench is dug too close to the tree.

While underground line installation can harm trees, there are steps one can take to protect roots during installation. The most important of these is a tree protection zone. Matheny and Clark (1994) recommend establishing tree protection zones using four steps modified from the British Standards Institute (2012):

1. Evaluate the species tolerance.
2. Identify tree age (young, mature, or overmature).
3. Using Table 4.5, find the distance from the trunk that should be protected based on trunk diameter.
4. Calculate the optimum radius for the tree protection zone.

Fazio (1999) suggests several ways to honor tree protection zones:

- Do not trench, pile backfill or debris, or park or drive equipment inside the tree protection zone.
- In open areas and if practical, curve the trench as far as possible around the tree, but at least outside the tree protection zone.
- Store excavated soil on the opposite side of the trench from the tree.
- Look for routes that avoid tree protection zones.
- If roots are severed, those over 2 inches (5 cm) in diameter should be cut flush on the tree side of the trench.
- Trenches should be backfilled quickly and the trench watered to prevent the exposed roots from desiccating.
- If there is no alternative to routing the line through a tree protection zone, tunnel under the zone. Tunnels should be at least 2 feet (0.6 m) deep, but preferably 3 feet (1 m) under trees fewer than 12 inches (30 cm) in diameter, and 4 feet (1.3 m) under larger trees.

Table 4.4 Engineering and construction alternatives to line-clearance tree work.

ADVANTAGES	DISADVANTAGES
Underground	
• Often eliminates tree-caused outages • Can dramatically improve reliability • Often preferred by electric customers	• Expensive • Competes with many other underground utilities (water, etc.) for limited space • Installation requires high restoration costs, particularly on established landscapes • Underground cable lifetime can be limited, leading to disturbance to replace or upgrade • Underground outages are difficult to find and repair • Safety and reliability risks from dig-ins
Alley or wing arms (framing a crossarm and braces off to one side of a pole)	
• Can sometimes avoid a tree or trees that are the opposite side of the alley arm	• May require a down guy to compensate for offset mechanical loads on the pole • May be a temporary solution as trees will continue to grow toward the conductors
Compact construction framing	
• Can decrease clearance space required in tree crown	• Increases the likelihood of phase-to-phase outages
Covered overhead primary or tree wire	
• Wire covered with polyethylene provides some level of insulating, which can reduce tree-related outages • Consistent with conventional construction	• Is not dependable in minimizing outages • Does not reduce the amount of clearance required • Just enough insulation that protective relays may not sense a downed wire due to tree contact, resulting in risk of a live wire on the ground and exposing the public to electrical hazard
Hendrix spacer systems (combines covered with compact construction)	
• Combination of covered wires and bringing individual phase wires close together, reducing the crown area of a tree affected by the distribution lines • Conductors have thicker covering than tree wire, so they are better insulated	• Probably does not reduce the amount of clearance that is required • Just enough insulation that protective relays wire, so they are better insulated may not sense a downed wire due to tree contact with a resulting risk of a live wire on the ground, exposing the public to electrical hazard
Aerial cable systems	
• Fully-rated insulated cables • Nearly eliminates tree-caused outages	• Expensive • Requires robust structures • Wires still need to be protected from mechanical loading and abrasion

Adapted from Goodfellow 1995.

Table 4.5 Guidelines for optimal tree protection zones for trees of average to excellent vitality.

Species tolerance	Tree age*	Distance from trunk in feet per inch of dbh	Distance from trunk in centimeters per centimeter of dbh
Good	Young	0.5 ft/in dbh	6 cm/cm dbh
	Mature	0.75 ft/in dbh	9 cm/cm dbh
	Overmature	1 ft/in dbh	12 cm/cm dbh
Moderate	Young	0.75 ft/in dbh	9 cm/cm dbh
	Mature	1 ft/in dbh	12 cm/cm dbh
	Overmature	1.25 ft/in dbh	15 cm/cm dbh
Poor	Young	1 ft/in dbh	12 cm/cm
	Mature	1.25 ft/in dbh	15 cm/cm dbh
	Overmature	1.5 ft/in dbh	18 cm/cm dbh

*Tree age: young (<20% of life expectancy), mature (20–80% of life expectancy), overmature (>80% of life expectancy). Source: Matheny and Clark 1994. Examples of tree tolerance references can also be found in Matheny and Clark 1998 Appendix B and Hightshoe 1988.

IMPLEMENTING CONTROL METHODS

Laws and regulations governing integrated vegetation management practices and specifications written by vegetation managers must be followed. Integrated vegetation management control methods should be implemented at the appropriate time on regular work schedules to meet established objectives. Work should progress systematically and based on the progress of converting the right-of-way to compatible vegetation, using measures determined to be best for varying conditions at specific locations along a right-of-way. Some considerations used in developing schedules include the importance and type of line, vegetation clearances, workloads, growth rate of predominant vegetation, geography, accessibility, and in some cases, the time elapsed since the last scheduled work.

Clearances Following Work

The system operator should establish and document appropriate clearance distances or vegetation heights to be achieved at the time of work. Following work, vegetation on the right-of-way should consist of a height and species mix that meets management objectives: reducing electrical and gas safety and service-reliability threats until the next scheduled work, protecting the environment, and controlling costs.

The best practice is to remove incompatible trees, encourage compatible vegetation, and ensure—through ongoing monitoring and maintenance—that trees do not become established in these areas so no opportunity develops that would compromise the utility facility.

Debris Disposal

Debris, such as logs and slash, that results from IVM operations should be handled in a manner consistent with adjoining land use, terrain, aesthetics, wildlife habitat, and fire risk. Logs may be recoverable for firewood or timber products, and are often best left for property owner use or as wildlife habitat. They can also be chipped and either blown on site or hauled away or mulched with a mower. Brush and slash can be placed into piles, windrowed along rights-of-way edges, or lopped and scattered. Some jurisdictions may limit the height and length of slash piles. Slash or logs should not be placed below the high-water mark of streams or other bodies of water, unless to comply with a directive from a competent authority. Logs should not be moved from the work site if they are likely to be infested with an epidemic-causing disease (such as oak wilt or Dutch elm disease) or insect pest (such as emerald ash borer). Dead standing timber that cannot strike the line or violate mandated minimum clearance requirements can be left as wildlife habitat, where appropriate (e.g., in remote areas or in wildlife management areas). Logs might be decked and firewood diameter stems left on site for the property owner to recover.

MONITORING TREATMENT AND QUALITY ASSURANCE

An effective program must have documented processes to evaluate results. Evaluations can not only involve quality assurance while work is underway, but also after it is completed. Monitoring for quality assurance should begin early to correct any possible miscommunication or misunderstanding on the part of crew members. Early and consistent observation and evaluation also provides an opportunity to modify the plan, if need be, in time for a successful outcome.

Utility vegetation management programs should have systems and procedures in place for documenting and verifying that vegetation management work was completed to specifications. Post-control reviews can be comprehensive or based on a statistically representative sample. The results should be compared to objectives, referencing the baseline surveys completed earlier in the planning process. A review of environmental, customer, archeological, or other outcomes may also be necessary, along with property owner and stakeholder surveys. This final review points back to the first step and the planning process begins again.

Records are necessary for quality assurance and future planning. The type of information needed is best determined by the utility vegetation manager. Relevant data commonly includes details on land ownership, the date of pre-notification, and access. Records should be digitized (preferably on a GIS platform such as described in Chapter 2) and reflect dates of communication, names of stakeholders—and the nature of discussions with them, including any commitments. Records should also be maintained on the type and voltage of line or pipeline capacity, along with work dates, methods, and location. Where appropriate, records should be maintained on threatened and endangered species and other considerations.

CHEMICALS

Chemicals can be instrumental in IVM implantation, particularly in the chemically-facilitated biological control of cover-type conversion. Their use can be controversial, and improper application can result in unintended, negative consequences. To be used safely, they must be handled according to label directions. Applicators are not only required to comply with label instructions, but also all other laws and regulations pertaining to chemical use. It is important for utility arborists to understand some fundamental principles of chemicals and their use. An overview of many of these concepts follows.

Labels

Applicators are required to follow directions provided on **labels**. For example, in the United States, the Environmental Protection Agency (EPA) labels chemicals after research and testing assures that applicators, consumers, and the environment are protected (Miller 1993). In this case, the label is the law, and information it contains must comply with a standard format. The ingredient statement provides the chemical and common name of the active ingredient and identifies other components, along with the percentage each comprises of the product. The chemical name is technical and describes the chemical structure, while the common name is simpler and easier to understand. In the United States, only chemical names recognized by the EPA may be used on a label. U.S. labels also provide EPA registration and establishment numbers. Among other details, establishment numbers identify where the chemical was packaged. Labels list compatibility with other chemicals, recommendations to prevent overdose, hazards to wildlife, first aid statements, storage, and disposal, as well as protective clothing and equipment advisories (Miller 1993). Finally, the label describes the approved uses for the pesticide, and directions for its safe application.

Selectivity

Herbicides can be selective or non-selective. **Selective herbicides** control specific kinds of plants when applied according to the label. For example, synthetic auxins are a class of selective herbicides that kill broadleaved plants, but do not harm grasses. By contrast, **non-selective herbicides** work against both broadleaved plants and grasses. Non-selective herbicides can be effective where a wide variety of target plant species are present, such as during initial clearing or reclaiming dense stands of invasive or other incompatible plants.

Application techniques can also be either selective or non-selective. Selective applications are used against specific plants or pockets of plants. Non-selective techniques target areas rather than individual plants (see the *Herbicide Application Methods* section). Non-selective use of non-selective herbicides eliminates all plants in the application area. Non-selective use of a selective herbicide controls treated plants that are sensitive to the herbicide, without differentiating between compatible or incompatible species. Selective use of either selective or non-selective herbicides would only control targeted vegetation. Selective application is preferable to non-selective treatment unless target vegetation density is high (Miller 2014).

Toxicity

Toxicity is a substance's ability to damage an organ system by disrupting biochemical pathways or enzymes in areas other than the point of contact. **Acute toxicity** results from a single exposure or exposure over a short period of time. **Chronic toxicity** is the delayed damaging effect resulting from repeated exposure to low amounts of a substance over a long period or a lifetime (Ottoboni 1997). Damage to the point of contact is **corrosiveness.**

Exposure is the amount of chemical reaching the body, while **dose** is the quantity absorbed into the body. Every chemical, even water, can be toxic in sufficiently high doses. The risk of harm by exposure to chemicals is a combination of toxicity, dose and the likelihood of exposure.

Labels provide toxicity information in four categories (Miller 1993), as shown in Table 4.6:

Category I Highly toxic
Category II Moderately toxic
Category III Slightly toxic
Category IV Relatively nontoxic

Table 4.6 Acute toxicity.

Category	Signal word	Oral LD_{50} mg/kg	Dermal LD_{50} mg/kg	Inhalation LC_{50} mg/l	Human exposure (approximate oral dose to kill an average person)
I Highly toxic	DANGER-POISON Skull and crossbones required	0–50	0–200	0–0.2	A few drops to a teaspoon (5 ml) or a few drops to the skin
II Moderately toxic	WARNING	50–500	200–2,000	0.2–2.0	Between a teaspoon (5 ml) and an ounce (30 ml)
III Slightly toxic	CAUTION	500–5,000	2,000–20,000	2.0–20	Over an ounce (30 ml) to a pint (500 ml) or a pound (0.45 kg)
IV Relatively nontoxic	CAUTION to none	>5,000	>20,000	>20	Over a pint (500 ml) or a pound (0.45 kg)

Source: 40 CFR 156.62 and 40 CFR 156.64.

Most chemicals used in vegetation management are Category IV, relatively nontoxic (Senseman 2007).

Pesticides may enter the body through several different pathways. Dermal exposure can occur when wet, dry, or gaseous forms of pesticides contact the skin or other body surface, and are absorbed. The eyes, eardrums, scalp, and groin absorb pesticides more readily than other areas. Oil-based formulations are generally more easily absorbed through the skin than water-soluble pesticides (Miller 1993). Oral exposure can occur from directly ingesting chemicals, although this is rare. More commonly, oral entry results when applicators do not wash their hands before eating or smoking following work. Inhalation is a third pathway. The largest inhaled fractions tend to stay on the surface of the throat and nasal passages, while smaller particles can be inhaled directly into the lungs.

Acute toxicity is determined through tests involving a large number of animals. Animals are divided into equal groups, each of which is exposed to a single dose of different chemical concentrations, and the animals are observed for two weeks. Tests are conducted for dermal, oral, and inhalation exposure. Oral doses are administered with a feeding tube. Dermal doses are topically applied for 24 hours; and inhalation subject animals are subjected to specific chemical concentrations in enclosed spaces for one hour. The oral or dermal lethal dose that kills half the animals within the two-week test period is known as the LD_{50}, or lethal dose to 50 percent of test animals. Lethal doses are measured in milligrams of substance for each kilogram of body weight. Inhalation concentration that kills half the subject animals in two weeks is the lethal concentration or LC_{50}. Lethal concentrations can also be measured in both milligrams per kilogram per body weight or parts per million. The lower the LD_{50} or LC_{50}, the more toxic the pesticide (Miller 1993; Ottoboni 1997).

A **poison** is a substance with high acute toxicity. In the United States, substances that have an oral LD_{50} lower than 50 mg per kg or a dermal LD_{50} lower than 200 mg per kg of body weight (Category I toxicity) are categorized as poisons. The labels of poisonous materials are marked with a skull and crossbones. By contrast, a chemical that has an oral LD_{50} of 5 grams per kilogram or more, or a dermal dose of 20 grams or more per kilogram of body weight is classified as relatively nontoxic. Remember, the overwhelming majority of chemicals used in integrated vegetation management fall into this category.

In the United States, the EPA sets acceptable thresholds for chronic exposure by taking the lowest dose of a substance that shows any detectable effect on any test animal over a long period of time (including a lifetime), and divides that level by 100, so, acceptable exposure thresholds have a 100-fold safety margin below the lowest concentration of chemical that has been demonstrated to cause any long-term damage to any animal (Ottoboni 1997). Moreover, label rates of chemicals are far below acceptable chronic exposure levels, so herbicides will not cause long-term health effects when used properly.

Effects from exposure can either be reversible or irreversible. Reversible effects are not permanent, and can be changed or remedied, while irreversible effects cannot. Irreversible effects include mutation, oncogenicity (carcinogenicity), and teratogenicity (causing birth defects). Chemicals are tested for their irreversible effects. Chemicals commonly used in vegetation management have not been shown to cause irreversible harm to mammals (Sensemen 2007).

Half-Life

Half-life is the amount of time it takes half of the quantity of a substance to dissipate. Half-life is important to understand because it indicates herbicide persistence. For example, 2, 4-D and picloram have half-lives of 10 days and 90 days, respectively. Three main factors influence chemical persistence: soil properties, material characteristics and climatic conditions. Half-lives for many common herbicides can be found in Senseman (2007).

- Soil properties influence chemical persistence, including physical, chemical, light, and microbial characteristics (Hager and Refsell 2008). Soil physical properties that influence chemical longevity are adsorption, leaching, and volatilization, which are dependent on texture and organic matter content. In general, high amounts of clay, organic matter, or both contribute to persistence because these particles bind chemicals, keeping them from being lost to volatilization and leaching. Soil chemical properties that affect herbicide half-life include pH, cation-exchange capacity, and the concentration of essential elements. Chemical types are not alike in their chemistry, so various pH levels affect them differently. Cation-exchange capacity can influence chemical characteristics because some cations can react with chemicals and cause them to be more or less persistent, while some chemicals are not influenced by pH in any way (Hager and Refsell 2008).

- Microbial properties include the type and population levels of microorganisms in the soil, which can break down many types of chemicals. Soil microorganisms and climatic conditions are interrelated as microorganism population and vitality can be affected by soil temperature, pH, oxygen levels, moisture, temperature, and the availability of essential elements. Optimal climactic conditions, such as appropriate soil moisture and temperature, promote microorganism vitality and fecundity, which accelerate herbicide breakdown (Hager and Refsell 2008).

- Chemical properties, such as water-solubility, soil adsorption, susceptibility to photo degradation, sensitivity to microbial breakdown, and vapor pressure influence persistence and also affect the half-life of chemicals. Soluble herbicides dissolve in the soil solution and become subject to leaching, particularly if they are not adsorbed on to soil particles. Moreover, herbicides with low vapor pressure are more likely to persist because they volatize less readily than other herbicides (Hager and Refsell 2008). More on herbicide properties can be found in Senseman (2007).

- Moisture, temperature, and sunlight are primary climatic factors that influence chemical degradation. High temperatures and moisture generally increase degradation because they promote microbial activity. By contrast, cool, dry conditions cause most herbicides and tree growth regulators to persist. Increased sunlight will accelerate breakdown of many chemicals (Senseman 2007; Hager and Resfell 2008).

Tree Growth Regulators

Flurprimidol and paclobutrazol are two common active ingredients in tree growth regulators. By slowing growth rates of some fast-growing species, TGRs can be helpful management tools where removal or cover-type conversion are prohibited or impractical, such as in urban forest applications. They have not been demonstrated to be economically effective on large-scale, rural transmission facilities; however, they have proven useful in locations like urban distribution lines.

Although widely used, the term *tree growth regulator* is inaccurate because these substances are not technically plant growth regulators. Growth regulators are natural substances that control plant growth and reproduction. Commercial products called TGRs inhibit growth regulator production, particularly gibberellins, which promote cell elongation, among other processes. When gibberellins are suppressed, cells in many tree species do not elongate, so growth is slowed. Treated trees often have stunted shoots, and in some cases, darker foliage. If over-treated, some species will have dwarfed leaves. TGRs have been shown to reduce utility pruning time up to 59 percent (Redding et al.; Mann et al. 1995).

Herbicides

Herbicides are specialized pesticides that control plants by interfering with specific botanical biochemical pathways. There are a variety of herbicides, each of which behave differently in the environment and in their effects on plants, depending on the formulation and characteristics of the active ingredient. While appropriate herbicide use reduces the need for future intervention, misuse can carry environmental risks due to drift, leaching, and volatilization.

When properly applied, herbicides are effective and efficient. They can enhance plant and wildlife diversity and minimize soil disturbance (Yahner and Hutnik 2004). Herbicides reduce long-term costs and maximize the benefits of tree and brush removal with proper plant selection for control. Herbicides can release flowering herbs and forbs when they are used against woody plants that dominate a space. Releasing flowering herbs and forbs improves habitat for pollinators and promotes wildlife by enhancing escape and nesting cover. Noxious or invasive weeds can also be controlled on utility rights-of-way with herbicide treatment if that is a desired objective (see the *Biological Control Methods* section) (Miller 2014).

Herbicide use can control individual plants that are prone to re-sprout or sucker after removal. When trees that re-sprout or sucker are removed without herbicide treatment, dense thickets develop, impeding access, swelling workloads, increasing costs, blocking lines-of-sight, and deteriorating wildlife habitat (Figure 4.25). Treating suckering plants allows early successional, compatible species to dominate the right-of-way and out-compete incompatible species, ultimately reducing work (Miller 2014).

Herbicides used in integrated vegetation management are only slightly toxic or relatively nontoxic because they do not disrupt animal biochemical processes. Rather, they obstruct uniquely botanical biochemical pathways. This botanical biochemical disruption is an herbicide's **mode of action**. Common types include: ALS or AHAS inhibitors, synthetic auxins, photosystem I inhibitors, photosystem II inhibitors, and EPSP inhibitors (Senseman 2007). The modes of action of these herbicides are described in Table 4.7.

Closed Chain of Custody

Herbicides have customarily been supplied in concentrated forms in non-returnable containers. The practice required open packages of concentrate to

Figure 4.25 Herbicide use controls plants that sucker after removal and is the first essential step in cover-type conversion.

Table 4.7 Mode of action of classes of commonly used integrated vegetation management herbicides.

Chemical class	Mode of action	Common IVM chemicals
ALS or AHAS inhibitors	Inhibit a key botanical enzyme (actolactate synthase [ALS] or actohydroxyacid synthase [AHAS]) in the biosynthesis of critical amino acids.	imazapyr, misulfuron-methyl, sulfometuron-methyl
Photosystem I inhibitors	Accept electrons from photosystem I, forming hydrogen peroxides, which react to form compounds that destroy membrane fatty acids and chlorophyll.	diquat, paraquat
Photosystem II inhibitors	Inhibit photosynthesis by blocking electron transport, stopping CO_2 fixation.	diuron, tebuthiuron
EPSP inhibitors	Inhibit the enzyme 5-enolpyruvylshikimate-3-phosphate (EPSP) synthase, which leads to depletion of proteins or disruption of biosynthetic pathways required for growth.	glyphosate
Synthetic auxins	Act similarly to natural auxins, although the precise mode of action is not well understood. They appear to affect cell wall plasticity and nucleic acid metabolism. They also stimulate uncontrolled cell division and plant growth, which results in vascular tissue destruction.	2,4-D, dicamba, picloram, triclopyr

Source: Senseman 2007.

be used on job sites, which carried the risk of mixing or measuring errors, spills, over-application, and unnecessary waste to landfills. However, improvements in chemistry and application methods have reduced the volume of herbicide solutions required for control. These developments have made possible a **closed chain of custody** concept that involves ready-to-use and diluted concentrate formulations in closed delivery systems, which minimizes the risks associated with typical herbicide use (Figure 4.26).

Returnable containers increase crew efficiency, as crew members can direct more of their efforts to application, rather than the mixing, loading, triple rinsing and finally, disposing of packaging required in conventional methodology (Goodfellow and Holt 2011). The Utility Arborist Association has published best management practices to advance applicator and environmental protection (Goodfellow and Holt 2011). This section is based on that publication.

The closed chain of custody concept covers herbicide shipping, distribution, storage and mixing and includes returning empty containers for refilling and reuse. It includes four critical elements (Goodfellow and Holt 2011):

- **Container cycle. Supply containers** that are returned, refilled, and reused
- **Integrity cycle.** Closed connections at the transfer points between supply containers, mix tank, and application equipment
- **Documentation cycle.** A container tracking system that establishes an auditable record documenting movement of herbicides and containers
- **Herbicide cycle:** Use of custom blends containing the required active ingredient and adjuvants

Closed chain of custody connections minimize applicator exposure to herbicide concentrates, although personal protective equipment requirements are unchanged. There is potential exposure while connecting and disconnecting closed connections with supply containers, which may become pressurized when exposed to direct sunlight or subjected to changing temperatures (Goodfellow and Holt 2011).

Chapter 4 • Integrated Vegetation Management   **115**

Figure 4.26 Closed chain of custody involves ready-to-use and diluted concentrated herbicide formulations in closed delivery systems.

Figure 4.27 Returnable, reusable containers are made of chemically-resistant plastic with a service life of about 5 years or 30 return cycles.

Use of Returnable, Reusable Supply Containers

Returnable, reusable containers (R/R) are typically made of recyclable, translucent, chemically resistant plastic with a service life of roughly 5 years or 30 return cycles (Figure 4.27). They meet United Nations (UN) and Department of Transportation (DOT) standards, and class II container and hazardous waste requirements in the United States. Beginning in August 2011, returnable container regulations have directed that each returnable, reusable container have a unique identification number and a tamper-evident seal. The container has the following information (Goodfellow and Holt 2011):

- EPA labels for all the herbicide products it contains, including registration numbers
- The concentrations of the ingredients, including active ingredients, diluents, and adjuvants
- Specific lot or batch numbers
- Mixing/dilution instructions particular to the spray tank size in which it will be used (expressed in either the graduated units on the supply container or as a ratio)
- Identifiers for the utility and applicator. In the context of utility vegetation management, the owner is the utility and the applicator is the owner's agent

Containers are graduated in English and metric scales that enable accurate determination of the volume through the translucent sides (Goodfellow and Holt 2011).

Service Containers

Service containers are smaller-capacity receptacles that can be refilled by applicators from supply containers. They can be used within an applicator's operations for inner-company transport, but may not be used by anyone other than the applicator. Service containers typically have 2.5 to 5 gallon (approximately 10 to 20 liter) capacities, and should be sufficiently durable for repeated use. They are designed to hold the same amount of custom blend formulations as the application containers to minimize the need for spray crews to perform careful measuring in the field.

By law, one-way disposable containers are labeled "do not reuse" and may not be refilled and employed as service containers. Disposable packaging must be triple-rinsed and thrown out following its single use (Goodfellow and Holt 2011).

Copyright © 2018 by International Society of Arboriculture. All rights reserved.

Use of Closed Connections at Transfer Points

Each returnable, reusable supply container should have a closed mechanical, leak-proof interlocking valve or connection. There should be a closed connection at each transfer point where herbicide is moved from one container to another, or from a container to a tank. Closed interlocking valves are mechanical and leak-proof (Figure 4.28), although closed connections should be inspected regularly. Tamper-evident seals must be maintained on all returnable, reusable supply containers. These seals are usually attached to the closed connection valves.

There should also be a closed connection between the supply container and the applicator's equipment. This includes maintaining closed connections between the following applications (Goodfellow and Holt 2011):

- The supply container and spray tank associated with application equipment
- Supply containers and any mix tank used as an intermediary to application equipment spray tanks
- A mix tank (if used) and the application equipment spray tank

Ideally, there should be a closed connection between the mix tank (or supply container) and portable spray equipment (i.e., backpack sprayers) receiving an application-ready mixture. Service containers should have a closed connection valve or fitting that is used for filling. Transfer from service containers to application equipment may be open poured since there is usually no means of emptying service containers with closed connections. Pumps (either electrical or manual) may be used, although they need to have chemically resistant seals (Goodfellow and Holt 2011).

Measuring Quantities of Customer Blends

Since custom blends in ready-to-apply formulations do not require measuring and mixing before use, they offer the advantage of reducing or eliminating field measurement of herbicides and adjuvants in specific spray mixes. However, there are two cases where measuring could be necessary: (1) when supply containers have more mix than is necessary for the spray container, and (2) when less mix is needed than a tank holds.

The volume of any existing spray mixture should be calculated and accurately measured, considering not only the active ingredients, but also the diluents (i.e., water or oil) being added to the mix or spray tank. For cases where the returnable, reusable supply containers have more mix than is necessary for a 1:1 ratio between the container and spray tank, the application-ready mixture should be measured in full units using the graduated markings on the supply container. If a supply container holds a smaller amount of mix than a full returnable, reusable closed supply container holds, the exact amount required should be put into the tank from the supply container.

Closed System Measuring

Closed connections reduce applicator exposure and the chance of spills during measuring. Spill reduction is accomplished in several ways:

- Cone tanks, which are intermediate fixed volume, calibrated, transfer containers that allow an applicator to determine specific quantities of mix
- Graduated and calibrated flow transfer pumps with flow meters

Figure 4.28 Returnable, reusable containers have a closed, mechanical, leak-proof interlocking valve connection.

- Translucent graduated supply containers as well as mix or spray tanks, which allow applicators to determine liquid levels and the volumes of mix being transferred

Custom Blend Formulations

Returnable, reusable closed supply container use is optimized when a few basic core mixes are prescribed for specific programs and projects. While standardization is desirable, there may be a need to make changes in application-ready mixtures to accommodate variable weather or site conditions. Ready-to-use and ready-to-apply custom blends are preferable on low-volume basal and cut surface applications, while whole dilute concentrates are preferred for low-volume foliar, high-volume foliar, and aerial applications.

Paraffin oil-based foliar mix carriers (such as Thinvert®) function as both surfactants and drift control agents. These formulations require additional field agitation to assure the carrier and active ingredients are appropriately held in the suspension. The stability of the custom blend dilute concentrate formulations being supplied should be well established or demonstrated by having uniform color without visible layering. Dilute concentrates should not be held in inventory longer than a spray season, although they can generally be expected to remain stable for two years. Some granular or other dry forms are unstable and cannot be incorporated into customer blends.

Mixing

Concentrates and dilute concentrates should be added to the spray tank on the right-of-way job site. Spray equipment should also be filled from supply containers on site. Mixing should not be done at any location where water being used as diluent is acquired. Mixing should not occur near a water source, and a buffer should be left (100 feet [30 m]) from a body of water or wetland. Many products that do not go into solution, but occur as suspension in the mix (like dry flowables) may have to be added to the tank just prior to use. Supplemental active ingredient(s) may also be added to the mix through an open connection, should that be necessary. Open connection supplemental additions should be added off the right-of-way job site (Goodfellow and Holt 2011).

The proper mixing protocol is as follows:

- Half-fill the tank with diluents (water or oil)
- Add concentrate or diluted concentrate to the tank
- Add any supplemental adjuvants (surfactant, drift control agent, or other material)
- Add the remainder of the diluents to the correct final fill volume

Tracking and Recordkeeping

The movement of herbicides and containers through the closed chain of custody should be tracked through meticulous, auditable records. Data should be maintained regularly with the returnable, reusable supply container tracked by its unique identification number. Records should document the container status as full, empty, or partial.

Computerized application equipment is available with GIS, location, timestamp, and volumetric data-capturing capability. So, location, time, and the amount of herbicide is automatically loaded into a database. When combined with the mixture information on the container's barcode, precise reports can be automatically generated.

Electronic databases that can be accessed through the Internet (such as cloud-based applications) are optimal. Use of information applications like barcoding, scanners, and other field-enabled technologies provide automated container location tracking capability (Figure 4.29). Data can also include temperature, relative humidity, wind velocity and direction, and precipitation at the time of application. The database should be searchable, and routine summary reports should be available.

Inventory Management

A "just-in-time" approach to inventory management finds the proper balance of maintaining enough herbicide to meet short-term needs without purchasing, storing, and maintaining a surplus. Keeping more material than is required not only creates unnecessary costs, but can also lead to waste since some custom blends degrade over

Figure 4.29 Barcoding allows for tracking of container locations.

time and go bad if stored too long. Goodfellow and Holt (2011) recommend having no more than three weeks' worth of chemical supply and storing no more than half the material required by a program on hand at any given time, unless the excess will be used within a few days.

When evaluating inventory, managers should consider the amount and purpose of the product being ordered. For example, it is counterproductive to purchase and store foliar mixes during the dormant season when foliar applications aren't possible. Conversely, ready-to-apply basal mix inventories should be limited if no cut stump or basal applications are scheduled. The unique identification numbers assigned to R/R supply containers can be used to maintain inventory records and might include the status of the containers (full, partial, and empty).

Herbicides should be stored in secured locations that are inaccessible to the general public. This should not only apply to permanent storage areas, but also to full and partial containers in transit or assigned to crews or projects. The number of partially filled containers stored should be minimized (Goodfellow and Holt 2011).

Applicator Handling of Empty R/R Supply Containers

Returnable-reusable supply container handling requires a reasonable standard of care. They may not be used for purposes other than being returned to the original custom blender for refilling and reuse (Figure 4.30). Container closure integrity must be preserved for reuse. That means containers should not be punctured, have closed interlocking valves removed, or have their tamper-proof seals compromised.

The optimal goal should be to use and return returnable-reusable supply containers on a 30-day turnaround schedule, but no more than 60 days. Empty containers should not be held by applicators until a full pallet is stockpiled for shipment (Goodfellow and Holt 2011).

Refilling R/R Supply Containers by Custom Blenders

Custom blenders are solely responsible for inspecting and refilling supply containers. Consequently, it is illegal for them to produce diluted concentrates unless they are ordered for a specific owner for a particular purpose. Blenders pressure test supply containers to UN and DOT Class II regulations before refilling them. If containers are refilled with the same herbicide formulations with which they were originally filled, they do not need to be triple-rinsed. However, tanks, valves and other components must be cleaned thoroughly before refilling with a different blend. Newly filled containers should have their old labels removed and new content-specific labeling applied (regardless of whether or not the contents have changed). After its designed service life (5 years or 30 cycles), the container is cleaned, taken out of service and recycled.

Herbicide Recordkeeping

Herbicide records are required by law. Applicators should identify themselves, note the herbicide trade name, the active ingredient, and the EPA number in the United States. They also need to track the amount of herbicide applied, the location of the application, weather conditions at the time of treatment, how many trees or acres were treated, and other relevant factors.

Figure 4.30 Returnable, reusable supply container handling requires a reasonable standard of care. Containers should not be used for anything other than their intended purpose.

SUMMARY

IVM is an ecosystem-based, systematic process of modifying plant communities and controlling incompatible vegetation on a site, with an underlying philosophy derived from integrated pest management. IVM offers an array of environmentally sound, cost-effective methodologies from which managers select those most appropriate to meet their objectives. It is not a set of rigid prescriptions based upon set time periods, repeated unselective mowing, or broadcast spraying across entire rights-of-way widths without the objective of establishing diverse, compatible plant communities.

Planning should be documented, methodical, and structured to achieve successful outcomes. Proper planning is cyclical, flexible, and ongoing. It allows managers to anticipate and correct problems before they arise. Plans should be based upon regular inspection of utility facilities and designed to protect the environment as well as archeological and cultural sites from damage during work. It uses information gained in the initial stages of planning to determine future strategies and outcomes.

IVM includes manual, mechanical, chemical, biological, and cultural methods. Managers should choose the option or options most suitable to their objectives, and more than one measure may be appropriate. Methods should promote a tree-resistant plant community wherever possible. Tree-resistant plant communities reduce long-term workloads and costs in time by out-competing incompatible plants. Chemical application must be made according to label directions. Herbicides and tree growth regulators disrupt botanical processes and commonly have low toxicity to animals, including humans. Different types of herbicide applications include spot, localized, broadcast, and aerial. Closed chain of custody best practices should be utilized.

Best practices work with nature rather than against it. Vegetation managers have a responsibility to protect the public and electrical grid by making every reasonable effort to keep vegetation away from utility facilities. Integrated vegetation management best management practices are enabling tools toward that objective. They not only minimize cost and environmental impact, but in some cases also enhance wildlife habitat.

CHAPTER 4 WORKBOOK

Fill in the Blank

1. The _____ _____ is the section of the transmission right-of-way directly under the wires and extending outward about 10 feet (3 m) on each side. The _____ _____ is the remainder of the active right-of-way.

2. A _____ is a substance that has an oral LD_{50} lower than 50 milligrams per kg or body weight, or a dermal LD_{50} lower than 200 mg per kg of body weight.

3. The target of IVM is _____ _____, including noxious weeds and invasive species, that pose potentially unacceptable economic, social, or environmental risks.

4. SMART objectives for integrated vegetation management are _____, _____, _____, _____, and _____.

5. Level 1 risk assessments review large populations of trees from a specified perspective to identify those that have an _____ or _____ likelihood of failure.

6. Vegetation, height, density, or other conditions that trigger specific control methods are called _____ _____.

7. For clearing dense vegetation during initial establishment or reclaiming neglected or overgrown rights-of-way, _____ control methods are most efficient and cost effective.

8. When properly selected and applied, herbicides can enhance _____ _____ _____ diversity and minimize _____ disturbance.

9. _____ toxicity results from a single exposure or exposure over a short period of time. _____ toxicity is the delayed damaging effect resulting from repeated exposure to low amounts of a substance, evolving over a long period.

10. Mixing of chemicals should be done with a buffer of at least ____ feet (___m) from a body of water or wetland.

11. _____ _____-_____ is a technique in which non-translocatable herbicides are applied to control specific branches growing toward utility facilities.

12. _____ _____ is the management of vegetation by establishing and conserving compatible, stable plant communities using natural competition, animals, insects, or pathogens.

13. The section of a utility transmission right-of-way under the wires and extending out both sides to a specified distance is called the _____ _____.

Multiple Choice

1. Which of the following steps in an integrated vegetation management program should be completed first?
    a. defining action thresholds
    b. evaluating the site
    c. selecting control methods
    d. setting objectives

2. LiDAR technology is precise, and is effective for
    a. documenting the distance between trees and conductors
    b. calculating fluctuations in the distance between vegetation and conductors under various environmental conditions
    c. identifying trees within striking or arcing distance of lines in the event of tree failure
    d. all of the above

3. Most herbicides used in vegetation management are in which toxicity category?
    a. Category I Highly toxic
    b. Category II Moderately toxic
    c. Category III Slightly toxic
    d. Category IV Relatively nontoxic

4. What are the three primary climatic factors that influence herbicide and tree growth regulator degradation?
    a. moisture, temperature, sunlight
    b. moisture, wind, sunlight
    c. temperature, wind, sunlight
    d. altitude, temperature, moisture

5. The NERC Transmission Vegetation Management Program requires that utilities in North America have transmission vegetation management programs in place, but it does not require them to be documented.
    a. True
    b. False

6. Which part of the body most readily absorbs pesticides?
    a. eyes
    b. feet
    c. hands
    d. forearms

7. How far should a tree protection zone be from a mature tree of moderate tolerance to disturbance?
   a. one inch (2.5 cm) per inch (2.5 cm) of diameter at breast height
   b. six inches (15.2 cm) per inch (2.5 cm) of diameter at breast height
   c. one foot (30 cm) per inch (2.5 cm) of diameter at breast height
   d. two feet (60 cm) per inch (2.5 cm) of diameter at breast height

8. A biological control used to provide a competitive advantage to short-growing, early successional plants is known as
   a. subsurface injection
   b. pollinator promotion
   c. cover-type conversion
   d. cut-stump technique

9. An example of a cultural control technique in a right-of-way is
   a. selective herbicide application
   b. grinding roots of suckering tree species
   c. relocating lines underground
   d. planting of food crops

10. Oil-based pesticides are less easily absorbed through the skin than water-soluble pesticides.
    a. True
    b. False

11. Compared to overhead electrical lines, a disadvantage of installing electrical lines underground is
    a. potential for increased land disturbance
    b. the increased installation cost
    c. outages are often more lengthy
    d. all of the above

12. Within buffer areas near bodies of water
    a. herbicides and pesticides cannot be used
    b. as much compatible vegetation should be retained as possible
    c. cultural control methods are typically the most effective
    d. local stakeholders should be contacted prior to scheduling work

CHALLENGE QUESTIONS

1. What are the four critical elements of the closed chain of custody concept?

2. Why are set time periods, such as cycle lengths, unsuitable action thresholds?

3. Explain why mechanical pruning is recommended only in rural and remote areas.

4. Outline the recommended methods for monitoring treatment and assuring quality within a utility vegetation management program.

5. Explain why site evaluations are critical for planning work and list some factors that should be taken into account during the assessment.

5

Electrical Knowledge

OBJECTIVES

- Communicate using appropriate electrical terminology.
- Explain the electrical system from powerhouse to customer.
- Describe basic functions of common electrical system hardware.
- Identify vegetation conditions that could cause service interruptions.
- Perform work around electrical hazards according to applicable regulations.

KEY TERMS

alternating current (AC)
aluminum conductor steel reinforced (ACSR)
ampere (amp, A)
back feed
cascading outage
circuit
conductivity
current
current surge
delta construction
distribution lines
distribution network
electrical potential
electromagnetic field
electromotive force
FAC-003
fault
ground wire
hertz (Hz)
impedance
inductance
insulator

interconnection reliability operating limit (IROL)
intermittent fault
interruption
lateral
lightning arrester
line sectionalizer
lockout
megawatt
minimum vegetation clearance distance (MVCD)
momentary interruption
North American Electric Reliability Corporation (NERC)
neutral wire
ohm (Ω)
operation
outage
overcurrent
phase (Ø)
primary voltage
resistance
secondary voltage

service drop
short circuit
sine wave
stator
step-down transformer
step-up transformer
substation
subtransmission lines
supervisory control and data acquisition (SCADA)
switchyard
taps
transformer
transient fault
transmission interconnect
triplex
turns ratio
volt (V)
voltage gradient
voltage ratio
voltage regulator
watt (W)

INTRODUCTION

Electricity is the only product that is produced, transported, delivered, and consumed in the same instant (Driggs 2005). The system that instantaneously produces and delivers electricity was designated by the National Academy of Engineering as the most significant engineering accomplishment of the 20th century. It is so reliable that people in the developed world have come to expect a flawless power supply (Schewe 2007), and they consider themselves wronged during outages (Oberbeck and Smart 2004). Yet, power outages happen, and trees cause a large percentage of those service **interruptions** (Russell 2011). Moreover, tree–power line conflicts can create problems that lead to electrical and fire safety risks.

Utility arborists are responsible for minimizing tree-related electrical conflicts. In addition, utility arboriculture demands an understanding of tree biology, pruning, tree risk assessment, and other traditional areas of arboricultural expertise. However, what sets utility arborists apart is the need for command of electrical principles, systems, safety, and many other utility-specific subjects. This chapter provides some essential electrical knowledge to help utility arborists meet their responsibilities successfully and safely.

ELECTRICAL FUNDAMENTALS

Electricity can be difficult to understand because it is invisible. It can help to think of electricity as a fluid, like water. Just as water flows downhill, electricity seeks the path of least resistance to ground. That path runs through conductive objects, and there may be more than one route. When electricity goes to ground, it dissipates in concentric "ripples," similar to those caused by a rock falling into calm water (Miller 2002). Eventually, electrical charges return to their source (Wellman 1959).

Voltage

A **volt** is a measure of electrical force, similar to pressure in the hydraulic model. For that reason, it is also called **electromotive force**. Volts are analogous to pounds per square inch or kilopascals in a hydraulic system. Electrical engineers use the letter V to symbolize voltage. Volts are often measured in thousands, represented by a k for kilo (the international term for one thousand). So, 1,000 volts is one kV, or one kilovolt. Voltage used by North American utilities generally ranges from 120 V for end users and up to 765 kV for the largest transmission lines (Wellman 1959); in Europe and other areas, voltage ranges from 220 V to 765 kV, mostly alternating current (AC).

Voltage by itself does no work. It has the potential for work, so it is also called **electrical potential**. It is possible to have a high voltage line, fully energized to thousands of volts, with no electricity running through it. To understand how that is possible, consider a high-pressure water hose, fully pressurized to 250 psi (1,700 kPa) with no water running through it because the valve is closed. An electrical line behaves much the same way. If nothing is drawing electricity from a line, the "valve" is closed, so no electricity is running through the wire. When electric appliances are turned on, the "valve" is opened, electricity begins to flow, and it "realizes its potential" by starting to work.

Amps

An **amp** or **ampere** is a measure of current. **Current** is the amount of electricity flowing. It is abbreviated with a capital A. Engineers also sometimes use the symbol I (for intensity) to indicate amperes. One amp is the charge carried by 6.25×10^{18} electrons moving past a given point in one second (Wellman 1959). Amps can be considered analogous to gallons or liters of water per minute in the hydraulic model. Amperage is variable on a particular circuit, depending on demand. As demand increases, amperage increases.

Watts

A **watt** is a measure of electric power. One watt is one amp flowing at one volt. Watts are calculated by multiplying volts and amps ($I \times V = W$). One-

and-one-third watts equal one horsepower (watts are named for James Watt, who is credited with inventing an improved steam engine). Watts are represented by W or P (for power). Electrical fixtures are commonly labeled with the wattage they draw (e.g., 20-watt light bulbs). Consumers' use of electricity is measured in kilowatts hours ([watts used × hours] / 1,000). A **megawatt** is a million watts. Power generation plants often produce many hundreds of megawatts of power (Whitaker 1999).

Resistance

Resistance is opposition to current flow. It is inherent in all electrical systems to some degree, and dissipates electricity proportionally to the square of the current (Schewe 2007). Resistance is measured in **ohms**, and designated by the letter R or the Greek symbol Ω. The unit is named for Georg Sikmon Ohm, who in 1820 related resistance to volts (V) and amps (I) in what is known as Ohm's law (Wellman 1959):

$$R = V / I$$

So, by definition one ohm is one volt "flowing" at one amp (or any number of volts at the same number of amps). Ohm's law can be rearranged to show that at constant resistance, increasing voltage increases amperage:

$$V / R = I$$

This equation demonstrates that as voltage increases, amperage has to increase for resistance to remain constant. Similar manipulations can be used to relate volts to watts, amps to ohms, watts to amps, and so on as depicted in the Ohm's law circle diagram (Figure 5.1).

Impedance

Impedance is often incorrectly used as a synonym for resistance on AC systems. In actuality, resistance constitutes roughly 90 percent of impedance, with the remainder comprising inductive and capacitive reactance. Inductive reactance is caused by a small amount of load lost whenever a current flows into a coil (such as those in transformers). Capacitive reactance is a charge that

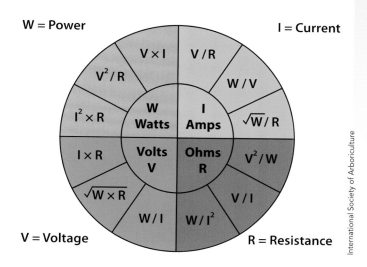

Figure 5.1 Ohm's law circle diagram.

builds up between long parallel conductors, which slightly hinders current. So, the sum in ohms of resistance, inductive reactance, and capacitive reactance equals total impedance in a circuit (Van Soelen 2006).

Conductivity

Conductivity is the capacity of a material to transmit electricity. Materials with high conductivity have low resistance to electricity, while poor conductors have high resistance. Poor conductors require greater voltage to accommodate electrical flow than good conductors. Aluminum and copper are good conductors, which is why they are often used for electrical wire. Fiberglass, glass, polymers, and porcelain are poor conductors, which make them suitable for insulators. Conductivity is the reciprocal of resistance.

Circuits

Electrical current completes a circle or **circuit**, to and from its source. Consequently, by definition, a complete electrical circuit needs a return path. The return path may be a neutral, ground, or other energized wire. A break in the circuit stops the flow of current (Van Soelen 2006). In North America, some utilities also refer to distribution circuits as feeders.

Outages, Overcurrents, Faults, and Current Surges

A power **outage** is a sustained service interruption (Russell 2011). **Overcurrents** are amperages in a conductor that are larger than those for which it is rated. They can be caused by lightning, faults, or improper design. A **fault** is a bypass of an intended conducting path. Fault current is the amperage of the resulting abnormal electrical flow. Current that is bypassing a designed conducting path is also called a **short circuit** (Whitaker 1999). "Shorts" can occur between a phase (energized conductor) and another phase, a phase and the system neutral, or a phase and the ground. These types of faults cause overcurrents (Van Soelen 2006). They can be direct, such as when two phase wires contact one another, or indirect, as might be caused by tree branches bridging two phase wires (Russell 2011). **Transient faults** affect the dielectrical properties of a system for an instant, and no longer exist after the power has been restored. **Intermittent faults** are repeated momentary interruptions in the same place due to the same cause (Institute of Electrical Engineers 1983), often due to sporadic equipment malfunction.

Current surges are inrushes of electricity. They are caused when electrical devices, like motors or transformers, are first turned on, at which time they can draw several times their normal operating electrical load. Current surges often occur when circuits are first re-energized following an outage.

Inductance

Inductance is electrification of a wire by passing it through a moving **electromagnetic field**. In 1831, James Faraday developed the law of electromagnetic induction when he established voltage in a coiled wire. Faraday theorized that lines of force flow out of the north pole into the south pole of a magnet and that these lines of force create electromagnetic fields. Lines flowing in the same direction relative to magnetic poles repel one another in the same way as similar magnetic poles. Faraday's theory is largely conjecture, but if magnets did not behave according to his assumptions, the foundation of electricity collapses and electrical systems would not work (Wellman 1959).

POWERHOUSE TO ELECTRICAL OUTLETS

Figure 5.2 provides a simplified diagram of a typical electrical system. In this basic diagram, electricity originates at generation plants, passes through transmission, subtransmission, distribution, then secondary voltage, and finally service lines to supply end users. Industrial customers are sometimes served by subtransmission lines. Table 5.1 shows common voltage classifications.

An energized conductor or wire in a circuit is called a **phase** (identified as Ø). Three phases are produced in field windings in generating facilities, each carried in a single wire. So, on many transmission or distribution circuits, there are three phases, often labeled A, B, C; X, Y, Z; or by color, depending on the utility.

Power Generation

Electricity is usually generated in power plants at distribution voltage, typically between 7.5 and 13.8 kV (IBEW/NECA 2006). In some cases, electricity flows directly from generation facilities into the distribution system. However, far more commonly, it is stepped up to transmission voltage to transmit it long distances to areas of need. Power plants are often large industrial facilities located in remote areas far from the end users in population centers. One reason generation facilities are built in rural areas is to locate them close to fuel sources, since it is more expensive to transport fuel like coal, oil, or natural gas than electricity.

System Control Centers

The nature of electricity as a commodity that is simultaneously manufactured, transported, distributed, delivered, and consumed presents logistical challenges for utilities. Most overcome those difficulties in part by coordinating their systems from control centers. Computerized equipment in control centers continually analyzes and balances

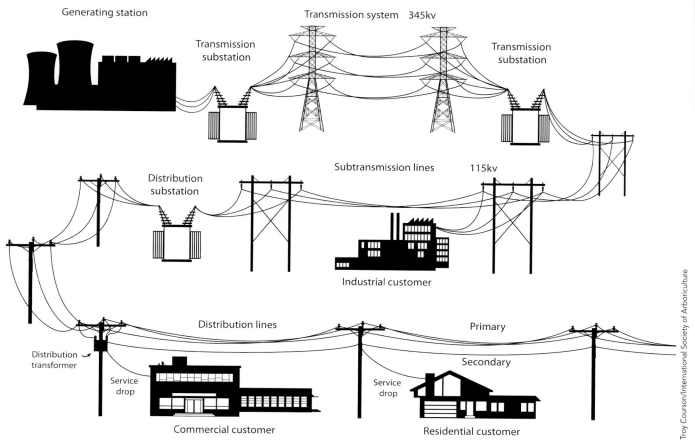

Figure 5.2 Electrical system layout.

Table 5.1 Common voltage classifications.	
Designation	Voltage
Extra-high transmission voltage	345 kV–765 kV
Transmission voltage	115 kV–230 kV
Subtransmission voltage	23 kV–115 kV
Distribution voltage	2.4 kV–23 kV
Secondary or customer voltage	120 V–240 V

Source: Abbott, Dubish, and Rooney 2005.

generation output and demand (IBEW/NECA 2006). Utility control centers use **supervisory control and data acquisition (SCADA)** systems. They automatically collect data and enable remote control switching operations (Van Soelen 2006).

Transmission

Transmission lines often originate at power plant **substations**. In many instances, the first lines coming out of power plants are stepped up to extra-high voltage—or at least 200 kV—and can run for hundreds of miles (km) from power plant substations to transmission substations. Most are AC. However, some of the highest voltage transmission lines are direct current (DC), which can be less expensive and have less electrical loss over long distances. Many are interconnected with main transmission lines from the same or other utilities. The resulting **transmission interconnect** provides efficiencies because electrical demand is rarely simultaneously high across broad geographic regions. It allows utilities to transfer electricity both within their systems and among other systems. It enables power delivery to areas of greatest demand, optimizing generating capacity, and avoiding the need to build expensive power plants that would only be used in periods of local peak usage.

The United States and southern Canada are divided into three interconnects: eastern, western, and Texas (Driggs 2005). Transmission lines that are part of each interconnect are those designated by the planning coordinator as elements of the **interconnection reliability operating limit (IROL)** or in the western interconnect as part of the major transfer path by Western Electric Coordinating Council (North American Electric Reliability Corporation 2008). The western United States provides an example of how interconnects provide efficiencies. Electrical demand historically peaks in the winter in the Pacific Northwest and during the summer months inland in the Rocky Mountain region. The interconnected transmission system allows Rocky Mountain fossil fuel plants to help meet high winter loads in the Pacific Northwest, while hydroelectric facilities in the Pacific Northwest supplement Rocky Mountain generation during peak summer months.

Transmission Substations and Switchyards

Transmission lines terminate at substations and switchyards. Substations change voltages, using transformers to step up or step down the voltage in lines (Figure 5.3). Switchyards enable utilities to route power through various circuits as needed. **Switchyards** may be designed strategically to compensate for portions of a system that are experiencing power failures. Switchyards protect circuits through disconnect switches, circuit breakers, relays, and communications systems (Van Soelen 2006).

Subtransmission

Subtransmission lines are often energized between 23 kV and 115 kV (Table 5.1), although some utilities have subtransmission lines as high as 230 kV or more. Unlike transmission lines, subtransmission lines are not interconnected. Rather, they link subtransmission and distribution substations to one another. Subtransmission lines may be built over distribution circuits (Figure 5.4). Several subtransmission lines may branch out from transmission substations to deliver power to industrial users and distribution substations (Figure 5.5). Subtransmission substations often feed the distribution system and sometimes directly

Figure 5.3 Transmission substation.

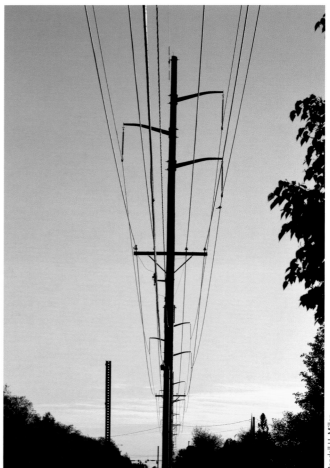

Figure 5.4 Subtransmission lines.

Figure 5.5 Distribution substation.

serve industrial customers. Generally, substations are served by multiple lines so they can depend on more than one source of energy supply.

Distribution

The **distribution network** is comprised of distribution substations, distribution circuits, secondary and service lines as well as other primary distribution equipment. The comparatively low potential in primary **distribution lines** is safer and easier to work with than transmission lines. However, the lower voltage comes at a price because its delivery range is constrained. For example, while transmission lines can run for hundreds of miles (km), a distribution 13.2 kV circuit is limited to about 30 miles (50 km) and 4 kV is only good for about 10 miles (16 km) before voltage loss (or voltage drop) caused by impedance in the lines can be problematic (Abbott et al. 2005).

The first lines out of substations are generally three phase. Single-phase lines are fed off of three-phase construction and are called taps, laterals, or feeders, depending on the utility. Single-phase circuits have only one wire conducting electricity (Figure 5.6) (Abbott et al. 2005).

Figure 5.6 Three-phase distribution line (right) and single-phase distribution line (left).

Distribution Substations

Distribution substations contain transformers that reduce subtransmission voltage to distribution **primary voltage**, usually between 2.4 kV and 23 kV (although some utilities have distribution voltage up to 35 kV). Several distribution circuits can radiate out from a given substation. Circuits originate at the terminals of a circuit breaker in distribution substations. Circuit breakers may have automatic reclosers.

Secondaries and Services

Secondary voltage for residential and business customers on North American distribution systems is 120 to 240 volts (Table 5.1). Distribution step-down transformers located outside homes and businesses are mounted on poles for overhead (Figure 5.7) and cement pads for underground lines (Figure 5.8). They reduce distribution to secondary voltage. The advantage of secondary voltage is that it is sufficiently low to be safely used for consumption. However, it has only enough force to drive electricity about 500 feet (152 m) before resistance in the lines compromises voltage levels.

Distribution Systems

Distribution systems are designed using a number of different strategies. Simple designs are less expensive, so they are preferred where practical. Complexity is sometimes needed for redundancy, which is required in areas of high customer demand or where there is a need for greater dependability (e.g., hospitals, emergency-response centers, or computer chip plants).

Radial Systems

Radial systems are the simplest distribution design. They have only one path of power flow (IBEW/NECA 2006). They have a single track of three-phase lines going from a substation, with three or single-phase laterals tapping off of them (Van Soelen 2006). While it is straightforward and inexpensive, the layout provides no backup source of power for customers, so the entire customer load beyond a fault is left without electricity during an outage.

A loop radial system is a common variation of a simple radial design. It essentially joins two radial circuits together with a normally open switch at their junction. The switch can be closed so parts of one circuit undergoing a power outage may be energized from the other. In that way, as few customers as possible are affected by the service interruption (IBEW/NECA 2006).

Figure 5.7 Distribution transformer on single-phase distribution Wye construction.

Figure 5.8 Distribution pad mount transformer.

Loop Primary and Primary Networks

A loop primary circuit is similar to a loop radial system except that it is entirely made up of one circuit that begins at a substation, makes a loop, and returns to the same substation. Lateral taps off the loop are generally radial (Van Soelen 2006). The advantage of the design is that it maintains power automatically when there is a fault in the line. Circuit breakers in the substation and automatic reclosers along the line work in conjunction to isolate a fault. Once the fault is isolated, relays on either side of the loop open to re-energize each side so only the distance between the protective devices flanking the fault is out of power. This maintains electricity to the longest line portion and the most possible customers during outages.

Primary networks are similar except that the loop involves multiple substations and feeders. It provides the maximum redundancy and protection from outages. It is also the most expensive strategy, so it is often limited to important commercial or industrial areas in densely populated areas, where electrical reliability is critical (Van Soelen 2006).

ELECTRICAL HARDWARE

Generators

Power generators convert mechanical energy to electrical energy using induction. Generators consist of large industrial magnets encased in wire windings that are known as field windings or field circuits (Figure 5.9). The field winding encasement is a **stator**. Magnets are spun on their axes, exposing the field windings to alternating positive and negative magnetic poles, which provide the movement necessary to induce electromotive force in the stationary coil. **Alternating current,** or **AC,** is created by the change in direction of the oscillating magnetic poles (Whitaker 1999). Alternating current periodically varies in amplitude in conjunction with the spinning magnets, intensifying from zero to a maximum value in a positive direction, back to zero and minimum value in the negative direction before returning to zero. The changing intensity and direction of alternating current is graphed in a **sine wave** (Figure 5.10)

Figure 5.9 Electric AC steam turbine generator.

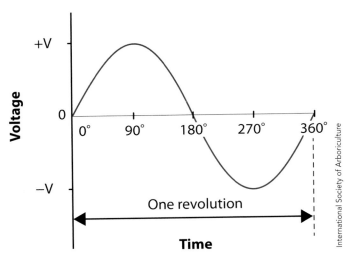

Figure 5.10 Single-phase sine wave.

(Wellman 1959). The time it takes for a magnet to complete a revolution is a cycle. The number of cycles per second is measured in **Hertz** or **Hz** (after the German physicist Heinrich Hertz who discovered electromagnetic waves in the 19th century). Smooth grid functioning requires electrical generation to be synchronized at the same frequency. In North America, electricity is generated at a constant speed of 60 Hz, or 60 cycles per second, ostensibly to synchronize electrical generation and timekeeping (Professional Training Systems 1979). In most other areas of the world, electricity is generated at a speed of 50 Hz (Van Soelen 2006).

A phase (Ø) is produced on a single winding (Whitaker 1999). Electricity is generated in three phases because stators are wound with three separate coils that are insulated from one another. Coils are set 120 degrees (one-third of a cycle) apart to produce three-phase voltage. The arrangement ensures against a break in electrical generation because there is no moment where all three phases are simultaneously at zero (Figure 5.11). Stator windings are connected to the electrical system in a closed circuit (Whitaker 1999), and each phase has its own line in the electrical system, which is why energized conductors are also called phases.

Fossil Fuel and Nuclear Steam Turbine Plants

Fuel-heated steam turbine plants generate the majority of electricity produced in the world. They use fossil (such as coal, natural gas, or oil) or nuclear-fission fuels. Uranium is used in nuclear reactors because it is energy dense. A specific w eight of enriched uranium contains nearly three million times the energy of the same weight of coal (Herda and Madden 1991). The fuel heats water to create steam, which is directed under pressure to spin the industrial magnets in three-phase generators to produce electrical energy (Whitaker 1999). Steam in fuel-heated boilers operates between 2,500 and 3,500 pounds per square inch (17 to 24 MPa) depending on design (Dorf 1993). Steam turbine generators are high-speed units that usually operate at 3,600 revolutions per minute to deliver 60 Hz output frequency (North America) through reduction gears (Whitaker 1999).

Fuel-heated steam turbine plants have advantages and disadvantages. Among their advantages is that they are reliable and provide a base load of electricity at a constant frequency, which is easily synchronized with other steam turbines in the electrical grid. The disadvantage of fossil fuel plants is

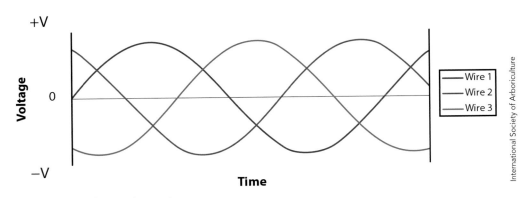

Figure 5.11 Three-phase sine wave.

that they contribute to climate change (Gore 2000; Evans 2007). Nuclear power plants don't contribute to climate change, so they are increasingly advocated as desirable sources of electrical generation (Evans 2007). Nuclear plants are increasingly subject to escalating costs due to greater complexity, inexperience in the nuclear power plant construction sector and stricter safety requirements in the wake of the Fukushima meltdown (Stapczynski 2017). However, many people have questioned nuclear safety for years, particularly in light of serious incidents at Three Mile Island in the United States, Chernobyl in the Ukraine, as well as more recently in Fukushima, Japan. In addition, authorities are concerned that processed fuel used for nuclear power generation may lead to nuclear weapons proliferation or terrorist use of radioactive waste for "dirty bombs" (Evans 2007). Disposal of radioactive waste also presents technical and political problems since its half-life can extend to millions of years. Attempting to keep it safe for such a long time seems unreasonable to critics (Glasstone and Jordan 1980; Evans 2007).

Gas or Combustion Turbines

Gas turbine generation is accomplished by routing hot exhaust gases from burning natural gas or oil directly to high-pressure combustion chambers where the gasses expand and spin turbines. The principle is similar to jet engine function. An improved design, known as a combined-cycle system, captures heat from the exhaust and uses it to create steam to drive a second turbine (Van Soelen 2006).

Renewable Generation Plants

Renewable energy sources are primarily inexhaustible. They include hydroelectric, wind, solar, geothermal, and ocean current, among other technologies (Evans 2007). This section covers the most common renewable energy sources, along with some advantages and disadvantages of each.

Hydroelectric Generation

Hydroelectric generation is the most common source of renewable generation. It is relatively inexpensive, simple, and dependably delivers base loads of energy at a constant frequency. It is a well-established technology used around the world. For example, two-thirds of electric generation in Canada and three-quarters in Brazil is hydroelectric. Hydroelectric facilities can be as large as 700 megawatts and create no greenhouse gasses or other air pollution (Evans 2007).

Hydroelectric generating facilities are simple compared to fuel-heated steam turbine plants. Hydroelectric turbines are mounted so they can be rotated directly by flowing water. The energy needed to spin the turbines is produced by a water head, which builds behind dams. The resulting pressure is released to drive turbines. Hydroelectric turbines are low-speed units, usually operated in North America between 120 and 900 rounds per minute to provide 60 Hz through reduction gears.

Some hydroelectric facilities are designed to temporarily store electricity. The design concept is known as pump storage. Storage is possible because electric generators are essentially electric motors running in reverse (Schewe 2007). During periods of low demand (e.g., overnight), generators are electrified from other energy sources and reversed to drive pumps to move water from a lower to an upper reservoir. When demand is high, electricity is recovered by releasing water back through the generators (Figure 5.12). An example of this type of facility is the Tennessee Valley Authority's Raccoon Mountain Generating Plant (Tennessee Valley Authority 2017).

A disadvantage of hydropower is that its capacity may be limited in drought years. Moreover, since it also requires damming of rivers, it can negatively affect migrating fish, and the resulting backwater often floods large areas, altering broad ecosystems. To reverse the environmental impacts caused by hydroelectric facilities, there are efforts to decommission and remove some generation dams (Rymer 2008).

Wind Energy

Wind energy is the second most significant source of renewable energy, behind hydropower (Evans 2007). Its power plants are referred to as wind farms or wind parks (Gipe 1995). Wind generators are also called turbines, and there are many different types (Figure 5.13). The largest wind turbines have a maximum capacity of 6 megawatts,

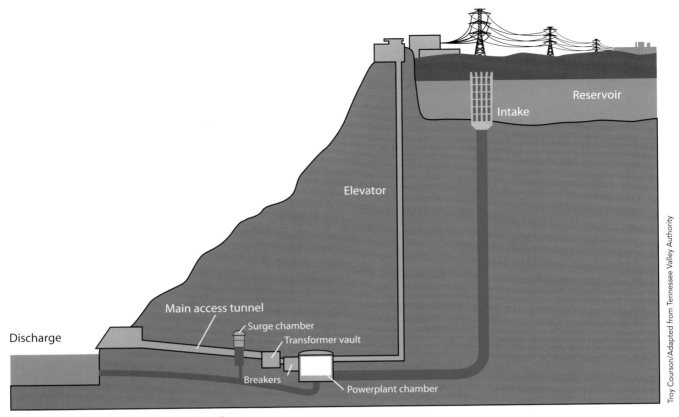

Figure 5.12 Pumped-storage plant diagram.

Figure 5.13 Wind turbines.

although that peak output depends on optimal wind speeds (Evans 2007).

Wind farms are constructed in areas with dependable, high-average wind speed (Evans 2007). Turbine blades are aerodynamically designed to maximize performance and create lift. The lift generates the pressure differential needed to rotate the blades (Patel 2006).

While wind is an inexhaustible energy source that doesn't produce greenhouse gasses, it has disadvantages. One of the main challenges of wind energy is that wind farms tend to be located far from load centers, and they are constrained by access to the transmission system. Furthermore, wind speed can be intermittent, which can make output undependable. Further, optimal wind speeds may or may not coincide with periods of peak demand (Evans 2007). The variable and unpredictable nature of wind velocity presents technical problems in balancing wind generation with demand and synchronizing it with the rest of the grid.

Aesthetic concerns have led many to oppose wind power plant construction (Gipe 1995; Schewe 2007). Bird mortality has been attributed to collisions with the turbines (Gipe 1995). Bats have also been killed by wind generators (Horn et al. 2006). Recent research indicates that the majority of wind-farm-related bat mortality may be caused by traumatic collision injury inflicted by the spinning blades (Rollins et al. 2012). However, some researchers argue that bats are too maneuverable to be killed by collisions with slow moving turbines. Rather, they attribute bat mortality to barotrauma. Barotrauma is caused by the abrupt air pressure changes on the back side of the turbine blades. These researchers argue that as bats pass by the turbines, the pressure changes cause lung hemorrhaging, killing the animals (Horn et al. 2006).

Solar Generation

Solar generation converts sunlight to electrical energy. The most common configuration for large-scale electric solar generation uses concentrating solar collectors (Evans 2007), which is a steam turbine design. Solar collectors use an array of mirrors over a large area to focus light on a receiver tower that generates sufficient heat to melt salt (Figure 5.14). The molten salt transfers heat to create the steam that spins turbine generators (Patel

Figure 5.14 Large-scale solar generating facility showing concentrating solar collectors.

2006). The methodology increases efficiency since molten salt increases generation hours because it holds enough heat to produce steam well into the night. It also improves efficiency during overcast conditions (Evans 2007). The most powerful solar power plants generate over 500 megawatts of electricity (Mendelsohn et al. 2012).

Solar power generation produces no greenhouse gas emissions, and its maximum output often corresponds with peak demand during the hottest part of the day (Evans 2007). On the other hand, solar collectors have been implicated in killing birds that collide with the reflective collection surfaces (Gipe 1995).

Photovoltaic cells have been utilized to produce direct current electricity on a small scale, suitable for individual homes and businesses (Evans 2007). Photovoltaic systems utilize semiconductors to convert sunlight to direct current (Figure 5.15). The system carries a high cost compared to large-scale power generation, so for now their use is often limited to locations that are off the electrical grid (Patel 2006). However, recent advances in photovoltaic technologies suggest that they have the potential to reach cost parity with other sources of generation in the not-too-distant future (J.W. Goodfellow, personal communication).

Geothermal Energy

Geothermal power plants exploit either naturally occurring steam or the earth's heat to produce steam needed to drive generators. Geothermal power plants are constructed in geologically active regions, which are characterized by localized high temperatures found relatively close to the earth's surface, and may be accompanied by geysers or hot springs. The most straightforward geothermal design harnesses steam in underground reservoirs formed when water hits hot subsurface rocks. The reservoirs are tapped and the steam is directed to the generators through pipes (Evans 2007).

Unfortunately, there are few places where steam is readily available at pressures and temperatures sufficiently high to efficiently drive turbines. Technology known as flash-steam has been developed to expand geothermal power production to exothermal areas where temperatures are otherwise too low to exploit. Flash-steam involves loading hot liquid or wet steam in a low-pressure container. The lower pressure causes the water to vaporize (i.e., flash) into steam, which is captured to spin turbines (Evans 2007).

Figure 5.15 Photovoltaic cells.

Ocean Current Generation

Ocean wave and tidal power is a nascent technology, which some consider to have the potential to become a benign source of electricity generation. While both ocean waves and tides are variable, the timing of tides can be predicted well into the future (Bedard 2007). Some of the disadvantages of wave and tidal power generation include appearance and noise, potential reduction in wave height from wave energy conversion, alteration of marine habitat, and potential toxic releases (Minerals Management Service 2006).

There are a variety of strategies for converting ocean wave to electrical energy. They include point absorbers, oscillating water columns, overtopping terminators, and attenuators (Bedard 2007).

There are two types of tidal turbines: vertical-axis and horizontal-axis. Vertical turbines have an upright shaft with a magnetized base encased in a stator on the ocean floor. They have hydrodynamic wings that are spun by tidal action to drive the magnet. Horizontal turbines function in a fashion similar to wind turbines except that they are turned by tidal currents rather than wind (Bedard 2007).

Structures or Poles

There are several types of structure or pole constructions depending on their purpose, line voltage, topography, and other factors. In general, the taller and more robust the structure, the higher the voltage in the lines they support.

Transmission towers of the highest voltage lines (typically 345 kV and above) are often steel construction (Figure 5.16). Lower voltage transmission lines may be built on multiple wooden poles (H construction), as shown in Figure 5.17. Most distribution poles are wooden, although steel and fiberglass are also used. Wooden poles are branded with information, such as their producer, the plant where the pole was manufactured, the year and date of manufacture, species of timber, and preservative treatment, class, and length (Abbott et al. 2005). In addition, many utilities mark and tag their poles (Figures 5.18a and b).

Figure 5.16 Steel transmission structures.

Figure 5.17 Transmission H structures.

Figure 5.18 (a) Branded wooden pole. (b) Pole tag.

Wires

Energized Wires

Energized wires or conductors are usually made of either copper or aluminum. Years ago, conductors were predominantly copper. Steel was also used, but its conductivity is only one-tenth that of copper so it was less common (Whitaker 1999). Recently, aluminum has gained favor due to its relative light weight and low cost in spite of having only 60 to 80 percent of the conductivity of copper. Aluminum wire's light weight means lines can be built with fewer structures and longer spans than are needed for copper. It also provides utilities with the opportunity to use larger gauge wire, which provides less resistance and accommodates higher voltage than thinner conductors.

Aluminum's major disadvantage is its tendency to stretch when it is heated during high electrical loads and elevated ambient temperatures. To minimize stretching and add strength, aluminum wire is stranded around a steel core. The resulting construction is known as **aluminum conductor steel reinforced** (**ACSR**).

Most energized transmission and distribution primary wire is bare, although some distribution conductors are covered with thin weatherproofing. Weatherproofing does not provide insulation, and offers no greater degree of safety than bare wire. Many times, weatherproofing on older wires deteriorates to the point that it shreds and hangs off of the wire, which can create a misconception of increased electrical safety risk. Occasionally, coated overhead primary, known as tree wire, is used on distribution lines in densely forested areas to provide some protection from faults caused by branches. However, like weatherproofing, tree wire is not insulation, and no safer than uncovered wire. Hendrix spacers may be used in conjunction with aerial spacer cables (Goodfellow 1995). Triplex is commonly used in areas of high tree density in Brazil (P. Castro, personal communication).

Neutral or Ground Wires

Neutral wires connect an electrical circuit or electrical equipment to the earth or to some conducting body that serves in place of the earth. They function as safety devices that ensure there is no difference in potential and no shock hazard among non-current carrying parts anywhere in the system. **Ground wires** are also used to connect equipment grounds, pole attachments, and hardware to a common equipotential zero voltage plain. Under normal circumstances, neutral wires do not carry a charge. However, they can carry primary voltages when there is a ground fault. They work by providing a low-resistance fault return path to the power source, and help facilitate proper overcurrent device operation during a ground fault. Ground wires are used to connect the system neutral to the earth. There are typically many grounds per mile of overhead line.

Abbreviations of Common Tree Species Used for Poles in North America

DF = Douglas-fir
EC = *Eucalyptus*
SP = Southern pine
WC = Western red cedar

Distribution Phase-to-Phase and Phase-to-Neutral Voltages

Most multiphase distribution lines are built using Wye construction, which consists of three phases and a grounded or neutral wire. The phases have polarity, creating a voltage differential among them. The potential difference between any two phases is approximately 1.73 times higher than the voltage difference between a phase and the neutral wire. For example, phase-to-phase 4.1 kV would be 2.4 kV phase-to-ground (Abbott et al. 2005).

Typically, the neutral wire on Wye construction is located beneath the conductors, on a crossarm or above the conductors at the very top of the pole. In the top position, it doubles as lightning protection, and is often called a static or shield wire. Phases are mounted on insulators while neutral lines usually are not.

Delta construction has three phases, but no common neutral wire, so there is no fixed difference between phase-to-phase and phase-to-ground voltages (Abbott et al. 2005). Workers should be alert to the possibility of delta construction to avoid mistakenly assuming that one of the lines is neutral. There are rare cases of delta construction on a single pole, where a primary wire is placed in a low position, similar to where one would expect a neutral wire on single-phase Wye construction. If workers fail to respect the low wire as potentially energized and do not maintain appropriate minimum approach clearances, they could be injured, maimed, or killed. This is one of many reasons to treat all wires as if they are energized.

Secondary and Services

Distribution primary wires carry electricity at voltages that are too high for most residential and commercial customers. So, distribution transformers reduce voltage to practical levels and deliver it via the secondaries or services to customers. Secondary wires typically run pole to pole while services are strung from the last pole to the service entry point or meter. Secondary wires often run along alleys, rear lot lines, and streets to serve residences and businesses. Service wires or **service drops** run from secondary wires or distribution transformers to end users. Secondary and service wires are generally **triplex** wires, which have two insulated conductors and a neutral wrapped around a supportive cable (Figure 5.19). A single conductor typically carries 120 V, and the two together total 240 V in North America, 220 and 440 respectively in many other areas. Some older construction uses "open wire secondaries," which are uninsulated conductors and neutral wires.

Guy Wires

Guy wires are intended for support, rather than electrical conduction. They are strong, galvanized steel wires often used where strength is needed, such as where a line turns or dead-ends, in areas of high winds, or at other anticipated stress points. Guy wires may be anchored to another pole forming a span guy, or secured at ground level in a down guy. Down guys may be outfitted with visible covers (Figure 5.20). While guys are not intended to conduct electricity, they are conductive because they are made of steel. Guys can become accidentally energized by direct or indirect contact with a live wire as could happen if a primary distribution line broke and fell onto a guy, for example.

Telephone and Cable Television Wires

Telephone and cable television (CATV) wires are not part of the distribution system, but are often joint facilities on electrical utility infrastructure (Figure 5.21). Telephone and CATV wires are typically located beneath the energized phases and

Figure 5.19 Service drop.

Figure 5.20 Guy wires with high-visibility covers.

Figure 5.21 Telephone and cable television (CATV) lines.

Figure 5.22 Underground cable cross-section.

the neutral wire. These wires generally operate at very low (under 150 volts) or no voltage. However, they are conductive, and can be accidentally energized by contact with energized lines (Abbott et al. 2005).

Underground Cables

Primary electrical lines have been installed underground since the 1880s (IBEW / NECA 2006). They are gaining popularity particularly in urban areas, at least for distribution application. Their use may be restricted by congestion, other underground utilities, and economics (Whitaker 1999). In most cases, transmission underground construction is prohibitively expensive so its use is limited to specialized cases.

Underground cable consists of all energized phases and the neutral in an insulated covering (Figure 5.22). Primary cable is often buried 3 feet (1 m) and secondary 2 feet (0.6 m) deep. While underground wires avoid interference with branches, it is more expensive than overhead construction, and underground faults are difficult to locate and repair. Furthermore, underground cables can interfere with tree roots and have limited life spans.

Switches

Switches are used to manage the electrical network (Whitaker 1999). They can be operated to shift load or reroute electricity between circuits during outages (Abbott et al. 2005). They are manually operated and not designed for line protection. Disconnect switches are generally used as visual isolation devices on three-phase distribution and subtransmission lines (Figure 5.23) (Abbott et al. 2005). One of their advantages is that operators can see open blades showing that two circuits or circuit sections are disconnected. Oil switches are a type of disconnect switch that can smother arcs that might result when heavy electrical loads are interrupted. Airbrake switches (Figure 5.24) are another common design, often used where circuits meet. The air gap created when the switches are opened provides the insulation needed to break the connection.

Circuit Protection

The electrical system is protected from anomalies—such as overcurrents, surges, and faults—by a series of safety devices. Protective devices also

Figure 5.23 Disconnect switches.

Figure 5.24 Open airbrake switches.

safeguard customers' electrical equipment. Their activation interrupts power and is called an **operation**. Protective devices function to accomplish the following (Van Soelen 2006):

- Minimize the time the public is endangered by electrical safety risks due to fallen conductors or hazardous voltages
- Reduce the time electrical equipment is exposed to potentially damaging voltages or current
- Limit the number of customers out of power during outages by automatically isolating faulted portions of a circuit
- Open and re-energize a circuit during transient faults

Circuits unprotected from overcurrents may be subject to a variety of risks (Van Soelen 2006):

- Damaged transformers, regulators and conductors due to overheating
- System voltage collapse, potentially harming end user equipment
- Increased risk of electrical contact to the public in case of downed conductors or direct or indirect contact

The selection of protective devices requires engineering analysis to determine the appropriate level of short circuit or fault current. The protective device must be sufficiently sensitive to function in as short a time as possible while being sufficiently robust to withstand expected fault currents and extinguish any resulting arcs without damage to the system.

Transmission Protection

Transmission circuit breakers disconnect abnormalities from the system. They operate in response to three kinds of common irregularities: abnormally high amperage, abnormally low voltage, and unequal current in the three phases. Relays are low-voltage switches. In response to voltage either stopping or starting to pass through them they will send signals to circuit breaker control mechanisms to operate. Relays detect measurable faults and cause circuit breakers to open in time to prevent equipment damage (e.g., substation transformers), while at the same time coordinating with other protective equipment to isolate the fault to as small an area or customer base as possible. Breakers are set to operate within a set number of cycles. Considering that electricity is generated at 50 or 60 cycles per second, the first operation is often initiated within tenths of a second following a fault occurrence (Van Soelen 2006).

Transmission Protection Zones

Circuit breakers are strategically placed to protect specific equipment in the transmission system. The most common zones are the generator, transformer, bus, and lines. Since lines are the longest zone, they have the most exposure and are usually the most frequently disconnected section (Van Soelen 2006).

Distribution Protection

Distribution lines are strategically divided into zones of protection in a manner similar to transmission lines. However, their design goes a step further insofar as, unlike transmission lines, distribution conductors are usually divided into protection zones. The zones are intended to safeguard electrical equipment and minimize the portions of a circuit that are out of power (Dorf 1993). In general, the closer the fault to a substation, the higher the potential fault current and the more electrical customers are likely to be affected. Utilities place various protective devices throughout the system in an effort to limit the de-energized portion to as close as possible to anticipated sources of equipment failure. Devices farther from the substation are usually set to operate at lower amperages and shorter time intervals than those closer in. Sensitivity and time intervals are coordinated with other protective devices on the circuit so that faults cause those closest to a downstream fault to operate, isolating the short circuit. While the line beyond the fuse is out of service, the rest of the feeder can be re-energized by automatic protection devices. When used in conjunction with system configuration, protection systems minimize the impact of lost electrical supply to customers.

Examples of principles that drive protective strategies:

- A set of protective circuit breakers with automatic reclosers are located in substations. Since distribution feeders originate at substations, this is where the largest number of customers are potentially affected.
- A series of reclosers, sectionalizers, and cutouts are installed on the main three-phase lines. The main three-phase lines serve the remainder of the circuit, including all of the single-phase taps. So, these devices are second in importance insofar as protecting service for the most customers.
- Cutouts or reclosers are located where single-phase **laterals**, **taps**, or feeders extend off the three-phase primary distribution line and affect the third largest group of customers.
- Cutouts or reclosers along single-phase taps protect distribution transformers and associated secondary lines, and a smaller number of customers.

Circuit Breakers

Circuit breakers are the principal protective devices in distribution substations (Figure 5.25), just as they are in transmission substations. They are activated by relays that respond to faults and are designed to operate quickly, often within 10 cycles. An open circuit breaker de-energizes the entire distribution circuit, which can take hundreds or thousands of customers out of service (Whitaker 1999).

Automatic Line Reclosers

Once a circuit breaker opens, the entire circuit is out of service. While that protects the electrical system, it can unnecessarily cause long outages to a large number of customers. Some faults clear quickly, as may be the case with tree or animal contact, conductor clash, or other incidental interference. To ensure that such brief episodes do not cause unnecessarily long outages, engineers have developed automatic line reclosers (Figure 5.26), which are designed to reset quickly to give the cause of a fault an opportunity to clear. They may immediately recharge the circuit several times at adjustable time delays, often within seconds—two or three operations—before they open permanently. A permanent operation is called a **lockout** (Whitaker 1999). Lockouts require a lineman to manually reset the device. Automatic reclosers are installed in substations and on distribution lines, and often have current ratings between 10 and 600 amps.

Momentary interruptions or transient faults may result when automatic reclosers operate. A few decades ago, momentary interruptions were

Figure 5.25 Circuit breakers.

Figure 5.26 Automatic line recloser.

barely noticed by end users, as their only effect was flickering lights. However, now transient outages may disrupt digital clocks, computers, and other delicate electronic equipment, so utility customers are becoming increasingly sensitive to and intolerant of "momentaries."

Line Sectionalizers

Line sectionalizers isolate line sections or protective zones used to limit the number of customers out of service during outages (Whitaker 1999). They are mounted on distribution poles or crossarms (Figure 5.27). Line sectionalizers are coordinated with automatic reclosers but do not directly interrupt faults. Rather, they cut off current when an upstream recloser operates but before that recloser locks out. The process isolates a fault in the line beyond the sectionalizer (Whitaker 1999, Van Soelen 2006). To limit the number of customers affected by a fault, a circuit might have several sets of line sectionalizers. The first set out of a substation is typically set at the highest amperage, with lower ratings farther out where fewer customers are potentially affected by an outage (Abbott et al. 2005).

Figure 5.27 Line sectionalizer.

Fuses and Cutouts

Fuses are the simplest type of short circuit protection (Dorf 1993). They are installed on distribution lines to segregate both equipment and line sections, including taps. Common primary distribution fuses ratings are 8 A, 10 A, 25 A, 50 A, 75 A, and 100 A. Fuse sizes are selected to accommodate normal load but still protect the circuit during overcurrent conditions. Cutouts are fuses designed to operate above 600 V (Whitaker 1999). Typically, the farther out from the substation, the more sensitive the fuse (Abbott et al. 2005). Load break cutouts can also be operated as a form of switch to de-energize the equipment (typically a transformer) and line that the fuse is protecting. Fuses minimize the number of affected customers and eliminate the need for substation circuit breakers to open unnecessarily. When an overload or short circuit exceeds a fuse's limit, it will "blow" and fall open (Figures 5.28a and b), thereby isolating the fault and de-energizing the circuit forward of the blown fuse.

Lightning Arresters

Lightning (or surge) **arresters** are installed on distribution circuits in many areas to protect against lightning-caused voltage spikes and flashover faults that could damage equipment and interrupt

Figure 5.28 (a) Fuse and closed cutout. (b) Fuse and open cutouts showing gates hanging down, isolating the three primary phases on the right from those on the left. Note the second cutout from the left is closed; it serves the transformer and associated secondary line.

service (Figure 5.29). Arresters channel excessive lightning-caused current to ground, minimizing possible damage to utility equipment and customers' electric devices. During normal operation, lightning arresters function like insulators. However, when voltage spikes exceed the insulating capacity of the arrester, excess current is directed from a phase to a copper rod that is driven into the ground at the base of the pole.

Other Equipment

Insulators

Insulators provide support and safety by preventing short circuits and separating energized equipment from non-energized system components. They are made of poor-conducting material, such as porcelain, glass, ceramic, epoxy, polymer, or fiberglass. Different insulator designs and numbers of suspension units are used for particular purposes. The number, size, and length of insulators often indicate voltage (Table 5.2) (Abbott et al. 2005). For example, 500 kV and 765 kV transmission lines are bundled on V-strings (Figure 5.30), while distribution lines may have only one or two pin insulators (Figure 5.31).

Transformers

Transformers are critical links in the AC power system because they raise or lower voltage through inductance. Recall that Faraday's law of inductance holds that a wire coil can be electrified by passing it through a moving electric field, and the movement can be provided by alternating current (Wellman 1959). Transformers work using two sets of different sized wire coils. The alternating current from one coil transfers power to the other through induction. The two wire coils are each wrapped around a magnetic iron core bathed in non-electrolytic oil (Abbott et al. 2005).

Step-up transformers raise voltage, while **step-down transformers** lower it. Each transformer is technically both a step-up and step-down transformer, because induction works in either direction with alternating current. Whether a transformer is designated as step-up or step-down depends on the intended direction of the system. Requirements

Figure 5.29 Lightning arrester.

Table 5.2 Average number of suspension insulator units used per string for various line voltages.

Voltage	Number of suspension insulator units per string
13.2 kV	2
23 kV	2 or 3
34.5 kV	3 or 4
46 kV	3–5
69 kV	4–6
115 kV	7–9
138 kV	8–10
230 kV	12–16
345 kV	18–22
500 kV	24
765 kV	32

Reprinted, by permission, from Abbott, Dubish, and Rooney 2005. © ACRT, Inc.

Figure 5.30 V-string transmission insulators.

Figure 5.31 Distribution pin-type insulators.

for higher voltage beyond the device call for step-up transformers, while those for lower voltage demand step-down transformers. Electrical flow in an unintended direction is **back feed.** Electricity going into a transformer is designed as primary, while electricity leaving is secondary voltage. On step-up transformers, the primary voltage is lower than the secondary voltage. The opposite is the case for step-down transformers (Wellman 1959).

Each of the two coils in a transformer has a different number of turns. The ratio of one to the other is the **turns ratio,** while the proportion of primary to secondary voltage is the **voltage ratio**. The turns ratio and voltage ratio are identical. So, if the primary winding has ten times the number of turns as the secondary coil, the turns ratio is 10 to 1, and the voltage will be stepped down by a factor of 10 (Professional Training Systems 1979).

Distribution transformers are step-down units because they are designed to reduce the primary voltage to a lower, secondary voltage. Distribution transformers for overhead lines are mounted on poles, while those serving underground lines are commonly mounted on concrete pads (Figure 5.32). Step-down transformers are also commonly used to reduce transmission to subtransmission voltage, and to lower subtransmission to distribution voltage. Step-up transformers can be deployed to raise generation voltage to transmission voltage.

Power Consistency

The ability of transformers to increase voltage may give the mistaken impression that they generate electric power. However, voltage alterations don't change power, and wattage in a primary line has to be equal to the wattage in a secondary wire (or nearly equal as some energy is lost to heat and other factors) (Whitaker 1999). To understand how this works, recall that watts equal amps multiplied by volts. Assume an end user is drawing 50 amps at 120 volts, or 6,000 watts of power (50 amps × 120 volts = 6,000 watts) through a hypothetical perfectly efficient transformer. That same end user is only drawing 0.83 amps from a primary line energized to 7.2 kV (6,000 watts / 7,200 volts = 0.833 amps) on the primary side of the transformer. So, as the transformer alters voltage, the amperage changes proportionally (in this case, voltage is decreased and amperage increased by a factor of

Figure 5.32 Transmission transformers in a substation.

60), but power remains constant. This example also illustrates the concept of high voltage, low amperage and low voltage, high amperage lines.

Voltage Regulators

Voltage regulators are used whenever it is necessary to correct voltage irregularities. Inappropriate voltage variation can result from an array of operational conditions, including changing loads, voltage loss (e.g., due to the resistance in lines covering long distances) or other factors. The resulting over- or under-voltages can cause problems with electronic equipment. Voltage regulators remedy those problems by functioning like adjustable transformers, capable of either increasing or decreasing the circuit voltage as needed (Figure 5.33). Remember, that while they may increase or decrease voltage, they cannot manipulate electric power.

Capacitors

Capacitors temporarily hold and slowly release voltage. They might be thought of as short-lived batteries. The difference is that capacitors store energy in an electric field, while batteries do so chemically. They are used to boost voltage, eliminate sparks, smooth the flow of direct current and protect against momentary voltage deficiencies (Wellman 1959). The most common capacitor design consists of two parallel conductive plates separated by insulation. A charge slowly builds in the insulation and can be released slowly. Capacitors can be found in substations or mounted on power lines, usually one bank for each phase (Figure 5.34).

TREE-CAUSED ELECTRICAL SERVICE INTERRUPTIONS

Energized overhead supply conductors rely on air for insulation among them, or among energized conductors, a neutral conductor, or the ground. Trees can compromise that insulation and cause service interruptions (Russell 2011).

System Reliability Indices

Two prominent indices used to quantify reliability are SAIFI and SAIDI. They are calculated using simple division, with lower quotients indicating better reliability (Russell 2011). SAIFI is the System Average Interruption Frequency Index. It is a measure of the *number of outages* experienced by the average customer over a year (or any specific time period):

SAIFI = Total number of customer interruptions / Total number of customers

SAIDI is the System Average Interruption Duration Index. It measures the *number of minutes* the average customer is out of power over a year (or any specific time period):

SAIDI = Total duration of customer interruptions / Total number of customers

These reliability metrics help utilities determine the effectiveness of their practices. In 1998, the

Figure 5.33 Voltage regulators.

Figure 5.34 Three-phase capacitor bank.

Institute of Electrical and Electronics Engineers (IEEE) determined that North American utilities had a median annual SAIFI of 1.16 and SAIDI OF 88 minutes (Russell 2011). Reliability indices are variable among utilities because weather, environmental conditions, voltage level, facility age, the preponderance of underground lines, data collection accuracy, maintenance practices, and other factors affect each utility's system in different ways. The accuracy of data collection is crucial, as the results are only as good as the information upon which they are based. Russell (2011) reports that utilities often have a 25% outage data error rate.

The worst performing feeders may or may not indicate poor vegetation management conditions on a specific circuit, as outages can be caused by any number of factors besides trees. Vegetation managers can get a better understanding of how their programs are performing by using vegetation-specific SAIDI and SAIFI numbers. These indices capture the total duration or number of vegetation-caused outages, respectively, for a given time period. That requires knowing exactly how many outages are caused by trees, which demands accurate outage information. For many utilities, accurate information may not be available. If responders can't identify the cause of an outage, they often attribute the incident to vegetation, and that can skew the results. Reporting may be improved by assigning vegetation management representatives to field check reported tree-caused outages. Data can indicate the number of cases where the outage was due to grow-ins, overhang, branch or whole tree failure, or other parameters. Analysis of this information can be used to improve program management (Russell 2011).

Failure Modes

There are two types of vegetation-related failure modes: mechanical tear down and electrical short circuits. Either can happen first, or they can occur simultaneously (Russell 2011).

Mechanical Failure

Mechanical tear down failure is the most common cause of tree-related service interruptions (Goodfellow 2005). Pacific Power (Miller 2016) has found that 75 to 80 percent of the tree-related outages on their systems are due to failure of tree parts or whole trees. Approximately 70 to 86 percent of tear down incidents have been found to be caused by trees located outside utility rights-of-way, often at considerable distances. Mechanical tear down can be caused by large branches or entire trees failing and falling on or through electrical lines, poles, or other equipment. Tear down events tend to be associated with inclement weather, including high winds, heavy snow, or freezing rain. Unstable or structurally-flawed trees may also be responsible for tearing down power lines. Mechanical tear downs may or may not be accompanied by electrical short circuits, depending on local circumstances and system design (Russell 2011).

Electrical Short Circuits

Electrical short circuits can be produced when a tree or branch comes in contact with an energized conductor. The resulting short circuit can be a phase-to-phase, phase-to-neutral, or phase-to-ground fault. This typically happens when the tree part contacts wires or electrical equipment and creates a new, undesirable pathway for current flow. It can also occur when the tree pushes two wires together. Moreover, trees contacting power lines can erode and weaken conductors (Russell 2011).

Three tree and electrical system characteristics influence the likelihood of a tree-related service interruption: voltage gradient, stem diameter, and species.

The most important of the three attributes is **voltage gradient**. A voltage gradient is electrical potential over a specified distance. In the context of tree-caused electrical faults, it is determined by the voltage and spacing of the lines as well as stem diameter and species of tree (Goodfellow 2005). Greater branch diameter and closer phase spacing create higher gradients. Depending on tree species and branch diameter, the fault-no fault threshold ranges from 1 to 4 kV-per-foot (3.3 to 13.1 kV-per-meter); however, the probability of a voltage gradient of less than 2 kV-per-foot (6.6 kV per meter) causing an outage is almost zero (Russell 2011).

As a consequence, the closer the conductors or conductors and system neutral lines are spaced, the more vulnerable they are to tree-caused outages (Goodfellow 2005). For a distribution system, this means that three-phase compact construction is most exposed to tree-caused short circuits, and single-phase is the least vulnerable.

Tree-caused electrical short circuits take time to develop. When a tree stem lies across multiple phases or simultaneously contacts a phase and a neutral, electricity carbonizes its surface. In time, a low impedance carbon path may form through the branch that provides a short circuit conduit that can cause an outage (Figure 5.35) (Russell 2011).

Vegetation-caused faults resulting from trees growing into distribution conductors are rare. Succulent vegetation growing into power lines is desiccated at microamp levels, which means meristems are "burned off" and stop growing before they grow into high-voltage lines. This leads to "electrical trimming" that is accompanied by discolored leaves, which are commonly referred to as "burners," but they do not cause outages. A good example of this is shown in Figure 5.36 (Goodfellow 2005; Russell 2011). The dry, discolored biomass is actually less conductive than the moist vegetation that first contacts the line. However, short circuits can happen when lines sag into new growth, if the voltage gradient between wire and the stems they contact is sufficiently high (P. Appelt, personal communication).

Transmission Blackouts

While grow-in faults on distribution lines are rare, that is not the case on transmission lines. The elevated potential of transmission lines puts them at risk for outages caused by trees. In fact, faults resulting from trees contacting transmission lines have been problematic, and have initiated catastrophic outages affecting millions of people. An extreme case occurred on 14 August 2003 when 50 million people in eastern North America lost power for an extended period of time (Cieslewicz and Novembri 2004).

The electrical grid is susceptible to such widespread blackouts because of its interconnectivity.

Figure 5.35 The carbon path that led to an outage can be seen on this branch.

Figure 5.36 Tree that has been electrically "trimmed."

While interconnectivity is important insofar as it allows utilities to efficiently transmit electricity to areas of greatest need at a particular moment, it also potentially exposes the grid to failure. Pervasive high demand—such as that during widespread heat waves—stresses the grid, which can initiate system-wide instabilities. It is during such periods that tree-related outages have caused **cascading outages** that have driven an entire interconnect to collapse (Cieslewicz and Novembri 2004).

There has been a consistent scenario in cases of cascading tree-related blackouts. Widespread strong electrical demand during region-wide heat waves overloads transmission lines across an interconnect. The combination of high ambient temperatures and excessive electrical loads heats lines, causing them to stretch and sag into trees growing underneath (Figure 5.37). Tree contact causes faults that knocks lines out of service. The load from the disabled lines is instantaneously shunted to other already-overburdened transmission lines, stressing them until some of them are knocked out of service, which forces their load to other overtaxed lines, overwhelming them and so on, until runaway outages develop, power plants shut down and the entire interconnect collapses.

VEGETATION MANAGEMENT REGULATIONS

FAC-003

In response to the 14 August 2003 North American blackout, the Federal Energy Regulatory Commission (FERC) empowered the **North American Electric Reliability Corporation (NERC)** to develop a standard that requires North American utilities to have systematic programs that effectively prevent outages on the bulk transmission system. The result was **FAC-003** (NERC 2008). The standard's mission is to promote a defensive strategy for subject transmission lines to prevent vegetation-related outages that could lead to cascading. This standard covers transmission lines rated at 200 kV

Figure 5.37 Line sag scenario that can lead to cascading. (*Cieslewicz and Novembri 2004*)

and higher, or those designated as an element of an Interconnection Reliability Operation Limit (IROL) (mostly for eastern North America), or for western North America designated by the Western Electric Coordinating Council (WECC) as an element of a major transfer path (FERC 2008).

The standard has several requirements intended to improve electric transmission reliability:

- Vegetation is to be managed to keep it from growing inside flashover clearance distance for applicable lines (note that there are two nearly identical requirements with this provision. The first applicable to lines that are part of an IROL or Major WECC Transfer Path. The second to lines 200 kV and above that are not part of an IROL or Major WECC Transfer Path).
- Documented maintenance strategies, procedures, processes, or specifications should be developed to prevent vegetation from growing sufficiently close that it could cause flashovers. Those strategies have to take into consideration conductor dynamics, vegetation growth rates, control methods, and inspection frequency.
- Proper control centers must be notified without delay when vegetation conditions present an imminent danger of causing a flashover, and a corrective action plan to address that danger must be in place.
- Corrective actions must be established to prevent flashover distances from being violated due to work constraints, such as legal injunctions.
- Mandatory annual vegetation condition inspections must be established.
- Completion of annual work is needed to prevent flashovers.

The **minimum vegetation clearance distance (MVCD)** is a distance inside of which vegetation may not encroach. NERC has developed a technical reference to accompany the standard (NERC 2008).

The technical reference explains the standard's provisions, including a section on vegetation management programs, which emphasizes best management practices as described in ANSI A300, Part 7 (ANSI 2012) and ISA's *Best Management Practices: Integrated Vegetation Management* (Miller 2014).

National Electrical Safety Code

The National Electrical Safety Code (NESC) is currently an ANSI standard developed under the auspices of ANSI, with IEEE as the secretariat. The Code has been adopted by most states in the United States. Rule 218 of the National Electrical Safety Code (ANSI 2017d) is the most common vegetation management rule in the industry (CNUC 2010a).

Rule 218 reads:
A. General
1. Vegetation management should be performed around ungrounded supply conductors and communication lines as experience has shown to be necessary. Vegetation that may damage ungrounded supply conductors should be pruned or removed.

    NOTE 1: Factors to consider in determining the extent of vegetation management required include but are not limited to: line voltage class, species' growth rates and failure characteristics, right-of-way limitations, the vegetation's location in relation to the conductors, the potential combined movement of vegetation and conductors during routine winds, and sagging of conductors due to elevated temperatures or icing.

    NOTE 2: It is not practical to prevent all tree-conductor contacts on overhead lines.

2. Where pruning or removal is not practical, the conductor should be separated from the tree with suitable materials or devices to avoid conductor damage by abrasion and grounding of the circuit through the tree.

B. At line crossings, railroad crossings, limited-access highway crossings, or navigable waterways requiring crossing permits.

The crossing span and the adjoining span on each side of the crossing should be kept free from over-hanging or decayed trees or limbs that otherwise might fall into the line.

In 2007, the rule was changed from providing instruction to "trim" or remove trees that are "interfering" with ungrounded supply conductors to "prune" or remove trees that are "damaging" ungrounded supply conductors. The intent of the revision was to better convey the appropriate considerations of vegetation management. In particular, the word "interfere" had been interpreted in some quarters to mean that trees should be kept from contacting power lines. However, since incidental touching of leaves to energized conductors does not appreciably hamper the system, the Accredited Standards Committee never intended to prohibit all contact between vegetation and conductors (Clapp 2007). The 2017 standard goes farther, noting that vegetation management should be performed to the extent of what experience has shown to be necessary. Moreover, Note 2 has been added to say expressly that it is impractical to prevent all tree-conductor contacts.

Individual State Requirements

Several states in the United States have laws governing some aspects of utility vegetation management. The laws differ in their requirements. Some mandate inspections or property owner notification of vegetation management work. Others establish clearance requirements or prescribe cycle lengths or storm hardening. Seven states have notification requirements: Florida, Illinois, Missouri, New Jersey, Oklahoma, Virginia, and Wisconsin. Oregon and Washington have laws that protect utility vegetation management work from liability under certain conditions. Texas and Florida have storm hardening requirements. Three states have clearance requirements: California, Kansas, and Oregon. Kansas has a "no contact" rule, while California and Oregon mandate specific clearances (CNUC 2010a). In addition, the California Public Resources Code Section 4292 requires a 10-foot [3 m] vegetation-free cylinder to bare ground around any pole bearing equipment that can emit a spark during fire season in designated fire areas

(Nichols 1995). California, Illinois, Oregon, and Wyoming have regular vegetation condition inspections conducted by regulators (fire officials in the case of California) (CNUC 2010).

New Zealand

New Zealand has a comprehensive vegetation management law that places a share of responsibility for utility line clearance with tree owners (New Zealand Parliamentary Counsel Office 2003). Its purpose is to protect public safety and electrical supply, and it has four requirements:

1. Clearance requirements (Table 5.3)
2. Designated responsibility for "cutting or trimming" trees that encroach on electrical conductors
3. Assigning liability if the rules are violated
4. An established arbitration system to resolve disputes between the "works owners" (i.e., electric utilities) and tree owners

As shown in Table 5.3, there are two sets of clearance zones: the growth limit zone and the notice zone. The growth limit zone is a distance inside of which a tree may not encroach, even in high wind, freezing rain, and snow. The notice zone is one meter short of the growth limit zone. If utilities observe a tree within the "notice zone," the law requires them to issue the tree owner a "hazard warning notice," informing the tree owner that if the tree encroaches inside the growth limit zone, the utility will issue a "cut or trim" notice. The utility is responsible for the cost of removing or pruning the tree the first time it is needed, but thereafter tree owners must maintain the tree out of the growth limit zone at their expense. If they fail, the utility may prune the tree and bill the owner for the work. If a tree owner fails to comply with the law, they are responsible for any damage their tree may cause to a utility's facility. The law carries a provision for arbitration to settle disputes between the operator and the tree owner. It has teeth insofar as it carries up to a NZD $10,000 fine with an additional NZD $500 for each day the infraction continues. The fines apply equally to offenses committed by the tree owner and utility (New Zealand Parliamentary Counsel Office 2003).

Fire Rules and Regulations

Vegetation contacting power lines can begin to smolder, drop to the ground and start fires, which have the possibility of becoming catastrophic (Cieslewicz and Novembri 2004). Some jurisdictions have regulations intended to prevent fires from starting in this manner.

An example of the laws commonly adopted in the United States is the Urban-Wildland Interface Code (International Code Council 2006). The Urban-Wildland Interface Code can be adopted by any local jurisdiction and has the following key provisions (Cieslewicz and Novembri 2004):

- Prohibition against planting of trees that have the genetic capacity to grow within 10 feet (3 m) of any energized conductors
- Landscaping plan reviews to ensure the right tree in the right place are required
- Requirement for utilities to achieve a "maximum" amount of clearance at the time of work, and never allow vegetation to grow within 6 inches (15 cm) of distribution conductors

Table 5.3 New Zealand tree clearance requirements.

Voltage	Notice zone	Growth limit zone
66 kV+ (high voltage)	5.0 m (16.4 ft)	4.0 m (13.1 ft)
33–65 kV (high voltage)	3.5 m (11.5 ft)	2.5 m (8.2 ft)
11–32 kV (high voltage)	2.6 m (8.5 ft)	1.6 m (5.3 ft)
400 V / 230 V	1.5 m (4.9 ft)	0.5 m (1.6 ft)

Source: New Zealand Government Parliamentary Counsel Office. Electricity (Hazards from Trees) Regulations 2003.

- Requirement for utilities to clear vegetation from around certain poles as directed by local fire authorities
- Authorization for electric utilities to abate emergency tree and power line conflicts as defined in the code

SUMMARY

Utility arboriculture differs from other professions insofar as its primary objective is to reduce service reliability and electrical safety risks caused by vegetation. Consequently, utility arborists not only require command of customary arboricultural practices, but also electrical principles, systems, and safety.

Electricity is the only commodity that is produced, transported, delivered, and consumed at the same instant. Fundamentals of electricity include voltage, amps, watts, conductivity, resistance, inductance, and other elements. Electrical systems begin at power plants, where electricity is often "stepped up" by transformers to extra-high transmission or transmission voltage, which can run hundreds of miles to transmission substations, where voltage is reduced to subtransmission levels, delivered to distribution substations where it is reduced again to distribution voltage, carried via distribution lines to transformers, and finally delivered to customers through secondaries and services.

Electricity is generated by industrial magnets through inductance. Fuel-driven steam turbine power plants are the most common method of generation, using fossil and nuclear fuels. Hydroelectric, solar, wind, and geothermal plants are common renewable energy sources.

Switches are used to manage the distribution network. Overcurrent protective devices are strategically placed to minimize the portions of a distribution circuit that are out of power due to a fault. Protective devices include circuit breakers, automatic reclosers, line sectionalizers, fuses, and lightning arresters.

Transformers are critical links in the alternating current power system because they raise or lower voltage through inductance. Each transformer is capable of either increasing or decreasing voltage but is designated as either step-up or step-down, depending on its intended use in the system. While transformers manipulate voltage, they cannot increase or decrease electric power.

Trees cause outages through mechanical tear down and electrical short circuits. Mechanical tear down failures are the most common and are often associated with inclement weather. Electrical short circuits can be produced when a tree or branch comes in contact with two energized conductors, an energized conductor and the system neutral, or an energized conductor and the ground. The resulting short circuit can be a phase-to-phase, phase-to-neutral, or phase-to-ground. Voltage gradient, stem diameter, and species affect the likelihood of tree-related service interruptions.

Catastrophic transmission interconnect failures have occurred due to cascading outages initiated by trees. Notably, a blackout on 14 August 2003 left 50 million people without power in eastern North America. In response, the Federal Energy Regulatory Commission authorized the North American Electric Reliability Corporation to develop a vegetation management standard (FAC-003) to prevent vegetation-caused cascading outages on transmission lines of 200 kV and above, as well as lines specified by appropriate authorities as subject to the standard. Among other requirements, the standard directs utilities to have a vegetation management plan designed to prevent vegetation from growing within maximum flash distances from conductors, taking the maximum designed sag and sway of the conductors into account.

The National Electrical Safety Code is the most common vegetation management rule adopted by individual states in the United States. California and Oregon have strict clearance requirements. Florida and Texas have involved storm preparedness regulations. Finally, New Zealand has a national law where tree owners share financial responsibility for utility line clearance.

CHAPTER 5 WORKBOOK

Fill in the Blank

1. A/An _____ is a measure of electrical current.

2. Voltage used by North American utilities generally ranges from ____ V for end users (customers) to _____ kV for the largest transmission lines.

3. A _____ is a blocked current or a bypass of an intended conducting path.

4. Transmission lines terminate at _____, where voltages are changed using transformers to step up or step down the voltage in the lines.

5. Primary distribution lines have lower _____ than transmission lines, which is safer and easier to work with.

6. Fiberglass, glass, polymers, and porcelain are good _____; aluminum and copper are good _____.

7. _____ _____ is used on distribution lines in densely forested areas to provide some protection from faults caused by branches.

8. _____ _____, the principal protective devices in distribution substations, are designed to operate in as little as 10 cycles.

9. Electrical flow in an unintended direction is called _____ _____.

10. Three tree and electrical system characteristics that influence the likelihood of a tree-related service interruption are _____ _____, _____ _____, and _____.

11. Two indices used to quantify reliability are SAIFI and SAIDI, which measure the _____ and _____ of interruptions experienced by the average customer over a given period.

12. The most common cause of tree-related service interruptions is _____ _____ _____ failure.

13. The utility system's _____ allows the efficient transmission of electricity to areas of greatest need at a particular moment, which also makes the system more susceptible to _____.

14. The _____ _____ _____ _____, or MVCD, is a calculated distance inside of which vegetation may not encroach.

Multiple Choice

1. The capacity of a material to transmit electricity is referred to as
   a. electrical fault
   b. electromagnetic field
   c. conductivity
   d. inductance

2. A transformer that is intended to reduce voltage is known as a
   a. capacitor
   b. step-down transformer
   c. step-up transformer
   d. voltage regulator

3. Which of the following is *not* true of aluminum wire?
   a. it is light weight
   b. it is low cost
   c. it has more conductivity than copper
   d. it can span longer distances than copper

4. Which of the following is true of guy wires?
   a. they are intended for support, rather than electrical conduction
   b. they are made of steel and are therefore conductive
   c. they can become energized by direct or indirect contact with energized wire
   d. all of the above

5. Electricity seeks the path of least resistance to the ground.
   a. True
   b. False

6. Transmission circuit breakers operate in response to all of the following irregularities, *except*
   a. abnormally high amperage
   b. abnormally high voltage
   c. abnormally low voltage
   d. unequal current

CHALLENGE QUESTIONS

1. What is the difference between an automatic recloser and a line sectionalizer?

2. Describe the difference between a volt, a watt, and an ampere.

3. Explain the difference between Wye construction and delta construction, and its importance for utility line management.

4. Compare and contrast the Urban-Wildland Interface Code and the Uniform Fire Code with respect to their provisions for preventing fires from starting when vegetation contacts power lines.

5. List some of the protection features utilities use to protect their circuits, transmission systems, and distribution systems, and discuss some of the differences and similarities among the features.

6
Storm Preparation and Response

OBJECTIVES

- Assess the failure risk of vegetation given the type and intensity of typical storms in the region.
- Develop an integrated, detailed storm response plan.
- Mobilize a storm response appropriate for the extent of the storm damage.
- Coordinate logistics so that crews can safely and efficiently respond to the storm.
- Ensure that the needs of responding employees are met.
- Recognize special safety concerns during a storm response.
- Incorporate lessons learned after a storm response to make changes for future responses.

KEY TERMS

- Coriolis force
- cyclone
- derecho
- disaster
- disaster declaration
- emergency operations center (EOC)
- Enhanced Fujita Scale
- extratropical cyclone
- first responders
- freezing rain
- funnel cloud
- heavy, wet snow
- hurricane
- Incident Command System (ICS)
- large-scale response
- lever arm
- mutual assistance program
- pre-staging
- restoration
- restoration pruning
- Saffir-Simpson Scale
- sleet
- small-scale response
- staging area
- storm drill
- storm recovery
- storm response
- storm surge
- taper
- thunderstorm
- tornado
- track forecast cone
- tropical cyclone
- typhoon
- warning (storm)
- watch (storm)

INTRODUCTION

Storms have an enormous impact on trees, utility infrastructure, and the customers and communities that depend on utility services. When storms strike, demand for arboricultural services may outstrip local supply. This inherent vulnerability requires electric utilities to prepare response plans for the rapid **restoration** of services. This process can be viewed as a continuous, cyclical activity, as shown in Figure 6.1 (adapted from FEMA).

This chapter covers the steps involved in planning and preparing for effective **storm response** and **storm recovery**, including:

- Identifying sources of risk from storms, such as the different types and intensities of storms likely to occur in a given area, and likely patterns of tree damage and failure
- Anticipating possible storm damage and mitigating with proactive maintenance
- Planning and preparing for various types of storm responses
- Executing storm responses, including contingencies and logistical considerations
- Unique safety considerations during storm responses
- Strategies for handling news media, social media, and public relations
- Incorporating "lessons learned" into preparations for future responses

Besides damaging utilities and other property, tree failures during storms often block access for police, fire, utility, and other **first responders**. Utility arborists are often among the first to reach stricken areas and play a vital role in providing access and allowing restoration efforts to begin (Figure 6.2). In these circumstances, lives and livelihoods may be threatened, and utilities, municipalities, and other service providers incur considerable restoration costs.

The cost of restoring and repairing storm-damaged utility infrastructure can be significant, especially when combined with lost electrical sales in affected areas (Smith et al. 2014). But by far, it is the affected communities, including businesses and their employees, that bear the greatest financial impact of storm-caused electric service interruptions. Without electricity, most commerce grinds to a halt. The U.S. Department of Energy and the Congressional Research Service estimate that electric utility service interruptions caused by storms cost businesses and communities between 25 and 70 billion dollars annually (Campbell 2012).

Considering what is at stake, utilities, vegetation management contractors, and their suppliers must be prepared to mobilize and deploy personnel and equipment to provide a swift and well-coordinated storm response. This is true whether the response consists of one crew dispatched to remove a single fallen tree or involves thousands of personnel across an entire region. Worst-case scenarios must also be considered, both locally and in locations where personnel may be required to respond. It is also important to remember that storm work poses unique safety challenges that require proactive measures prior to and during responses (Chisholm n.d.).

Figure 6.1 The concept of disaster management as a continuous process is recognized around the world, and can be applied to any organization involved in disaster response.

Figure 6.2 Following storms, arborists' skills are essential for gaining access to stricken areas and restoring essential services.

Successful responses of any size, but especially large-scale responses, require open lines of communication, established chains of command, and flexibility as workloads and conditions change. These efforts may be coordinated through formalized regional storm emergency planning and response procedures that include government, utility, and other critical service providers and are consistent with Incident Command System (ICS) best practices and standards for emergency management. Well-planned and coordinated responses allow multiple organizations to work together to restore critical services to the communities and customers served by our industry.

RISK IDENTIFICATION

In the most general terms, the level of risk from any source can be determined by multiplying the probability of the occurrence of an event by the severity of the consequences (Smiley et al. 2017). For example, to assess the risk from storms, historic weather data can be used to determine the probability of various storm events occurring in a given area, and damage caused by other storms can be used to project the scale of damage from similar events.

Utility arborists should know the type and intensity of typical storms in their region, how these storms are likely to affect trees and infrastructure, and the steps necessary to properly prepare for and prevent or minimize damage. Risk can be better understood and preparations made more effective with an understanding of tree failure patterns during storms, and the knowledge that the condition of the trees present, including past maintenance practices, will affect system performance.

Effects of Storms on Trees

Loads

Trees are structures, and like all structures, they have breaking points. As trees grow, they continually adapt to support the load of their own mass against the constant pull of gravity, as well as added loads brought on by various types of precipitation and wind. Over time, trees also develop defects, such as leans, decay, girdling roots, and poor branch attachments. Damage and injury can also weaken tree structure, either due to natural causes or human activities. Storm conditions increase loads and stress which, when combined with defects, can lead to failure (Dunster et al. 2017). Most established trees can tolerate some increased loading, including typical rain, snow, and mild storm conditions. However, unusual or extreme conditions greatly increase likelihood of failure (Figure 6.3). Examples include:

- High winds
- Ice accumulating on twigs and branches
- Heavy downpours on dense foliage
- Wet snow, especially on trees with leaves on
- Combinations of wind and one or more of the above

Site Conditions

Site conditions also affect likelihood of tree failure. Trees that are exposed to the full force of wind, especially if they were previously protected, are more likely to fail. For example, a tree suddenly exposed to higher wind loads by the removal of an adjacent tree or trees may not be well adapted and could have a higher likelihood of failure. Likewise, root loss or damage caused by trenching, sidewalk replacement or other site disturbances may affect tree stability. When soils become saturated due to flooding or heavy rain, the friction between roots and soil is reduced, and the holding capacity of roots is compromised. These effects, when combined with wind and other added loads, increase the likelihood of failure (Dunster et al. 2017).

Tree and Branch Failure

Trees respond to the effects of loads with response growth to compensate for defects and inherently high-stress areas such as branch unions. The effectiveness of these adaptations depends on the tree species, as well as tree health and vigor, and the severity of defects. Loads are amplified or exacerbated by branch length (**lever arm**), the density of twigs or foliage, the degree of lean, lack of trunk or branch **taper**, wind exposure and other factors. The effects of wind are dampened by differences in the oscillation frequency of individual branches (James 2003). However, as wind speeds increase, the added load increases exponentially. Similar relationships exist with ice and snow loads. When accumulated loads exceed the stress tolerance of

Figure 6.3 Extreme conditions, such as accumulating ice, increase likelihood of tree failure.

the trunk, branch, or roots, failure of all or part of the tree is inevitable (Dunster et al. 2017).

Storm Types

Storms come in many forms, and the damage to trees varies depending on the type and intensity of the storm. Understanding how storms form, develop, and move is important in preparing a response strategy. Though many types of storms can cause severe damage to trees and utility infrastructure, different storms pose unique operational challenges.

National governments issue watches and warnings for potentially affected areas as storms become threatening or the probability of development increases. In general, a **watch** means that conditions are favorable for storms to occur, while a **warning** indicates that a storm is imminent or occurring and that immediate action is required (National Weather Service n.d.).

Cyclones

The term **cyclone** refers to any large area of circulating low atmospheric pressure, including tropical systems such as hurricanes and typhoons, and extratropical cyclones, which occur in higher latitudes (Figure 6.4). Due to the **Coriolis force**, cyclones rotate clockwise in the Southern Hemisphere and

Figure 6.4 A large extratropical cyclone off the west coast of North America. Note the counterclockwise pattern of circulation in the Northern Hemisphere.

Tropical cyclones form over warm ocean waters and in favorable conditions can quickly develop strong winds and heavy rains. Tropical cyclone frequency peaks during late summer and early autumn when waters are warmest (Figure 6.5). The most powerful tropical cyclones are known as **hurricanes** in the Atlantic and eastern Pacific, **typhoons** in the northwestern Pacific, and simply as tropical cyclones in the South Pacific and Indian Oceans (see *Areas Vulnerable to Tropical Cyclones* on p. 174) (Australian Government Bureau of Meteorology n.d.). The practice of naming these storms is managed by the World Meteorological Organization, an agency of the United Nations (World Meteorological Organization n.d.).

Various scales are used to rank the intensity of tropical cyclones. For example, Australia ranks storms from Category 1 ("negligible house damage") to Category 5 ("widespread destruction")

Figure 6.5 Tropical Cyclone Debbie off Queensland, Australia, 28 March 2017. Note the clockwise pattern of circulation in the Southern Hemisphere.

counterclockwise in the Northern Hemisphere, with winds increasing in intensity toward the center. Because of this pattern of rotation, the direction and intensity of winds change as cyclones approach and pass any given location. Cyclones are large, complex systems, often hundreds of miles (kilometers) across, and are often the underlying cause of various other kinds of storms. They are usually accompanied by one or more forms of precipitation (American Meteorological Society 2012). A cyclone is not the same as a tornado, which is a local phenomenon, although tornados often occur as a result of cyclones (Australian Government Bureau of Meteorology n.d.).

(Australian Government Bureau of Meteorology n.d.). Similar scales are used in other areas prone to these storms. In the Atlantic Basin, the **Saffir-Simpson Scale** ranks storms from Category 1 ("Very dangerous winds will produce some damage") to Category 5 ("Catastrophic damage will occur") (Table 6.1).

When tropical cyclones make landfall, damage from winds and flooding may be compounded by **storm surge**, which can inundate low-lying areas with seawater and debris and block access to damaged areas (Australian Government Bureau of Meteorology n.d.). The development of tropical cyclones is carefully monitored by both meteorologists and those who may be affected by a strike. Because their movements can be forecasted with some degree of accuracy, those in or near the forecasted storm track often have several days of notice prior to landfall.

Extratropical cyclones, as the name implies, occur at higher latitudes and often mark the boundary between different air masses that characterize changes in weather. They are frequently associated with severe weather, including high winds, heavy rain, winter storms, thunderstorms, and tornados. Extratropical cyclones are often larger than tropical cyclones but tend to have lower sustained wind velocities.

A tropical cyclone can become extratropical as it moves into higher latitudes. Superstorm Sandy was making such a transition as it made landfall on the eastern coast of North America in 2012. Sandy was unusual—and especially devastating—in that it combined high winds and storm surge typical of a hurricane with the large size and diverse conditions, including heavy mountain snows, more typical of an extratropical cyclone (National Oceanic and Atmospheric Administration Earth Observatory n.d.).

Thunderstorms

Thunderstorms are the result of rising areas of moist, unstable air (Figure 6.6). Heavy rain, lightning, thunder, hail, strong straight-line winds and

Table 6.1 The Saffir-Simpson Hurricane Wind Scale.

Category	Sustained winds			Types of damage due to hurricane winds
	mph	km/h	knots	
1	74–95	119–153	64–82	Some damage. Some tree branches snapped and some trees toppled. Extensive damage to power lines and poles, causing power outages lasting a few to several days.
2	96–110	154–177	83–95	Extensive damage. Many tree branches snapped and many trees uprooted, blocking numerous roads. Near-total power loss expected with outages lasting from several days to weeks.
3	111–129	178–208	96–112	Devastating damage. Many trees snapped or uprooted, blocking numerous roads. Electricity and water unavailable for several days to weeks.
4	130–156	209–251	113–136	Catastrophic damage. Most trees snapped or uprooted and utility poles downed. Residential areas isolated. Power outages last weeks to possibly months. Most of the area uninhabitable for weeks or months.
5	157+	252+	137+	Catastrophic damage. Nearly all trees snapped or uprooted and utility poles downed. Residential areas isolated. Power outages last weeks to possibly months. Most of the area uninhabitable for weeks or months.

Source: National Oceanic and Atmospheric Administration (www.nhc.noaa.gov/aboutsshws.php)

Figure 6.6 Thunderstorm building over the Florida Everglades.

tornados are often associated with thunderstorms (Scientific American 1999). Frequency and intensity vary regionally; for example, in North America, thunderstorms occur most frequently along the Gulf Coast and in Florida, U.S. However, severe thunderstorms occur more frequently in the central part of that continent. Straight-line winds in severe thunderstorms have been measured at over 120 mph (193 km/h) (Corfidi, Evans, and Johns n.d.). The likelihood of thunderstorm occurrence in any given area is often forecast days in advance; however, the precise path and intensity of individual storms is generally not known until hours or less prior to strike.

A **derecho** is a complex of thunderstorms that travels more than 240 miles (386 km), with wind speeds greater than 58 mph (93 km/h). Distances of more than 700 miles (1,100 km) and wind gusts over 100 mph (160 km/h) are not uncommon (Figure 6.7). For many inland areas, derechos are the most destructive storms in terms of widespread disruption, including tree damage, utility service interruptions, and cost of restoration (Corfidi et al. n.d.).

A **tornado** (also known as a twister) is a violently rotating column of air extending from a cloud to the earth (Figure 6.8) (Edwards 2016). In some areas, the term "cyclone" is incorrectly used to identify a tornado. Tornados form in certain types of severe thunderstorms and as tropical cyclones make landfall. A **funnel cloud** is a forming tornado; technically, it is not a tornado unless it reaches the ground.

Winds in tornados may exceed 200 mph (320 km/h) and can cause extreme damage where they

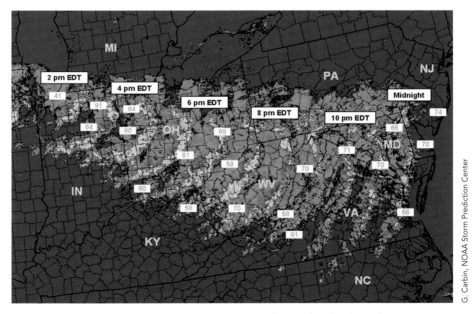

Figure 6.7 Composite radar image showing the path of a derecho across the eastern United States, 29–30 June, 2012.

Figure 6.8 An EF5 tornado near Moore, Oklahoma, U.S., 20 May 2013.

strike; however, damage is usually restricted to a relatively narrow corridor. The **Enhanced Fujita Scale** (EF Scale) is used to characterize tornados according to their estimated strength, from EF-0 to EF-5 (Table 6.2). EF ratings are based on observed damage following the storm, as it is not possible to measure the actual wind speeds in individual tornados.

Tornados occur in temperate regions around the world (Figure 6.9), but are most common in North America east of the Rocky Mountains, from southern Canada to the Gulf Coast, and eastward to the Atlantic Ocean, with the highest frequency from the south-central Great Plains eastward to the Appalachian Mountains. Some tornados are isolated, but widespread outbreaks are possible when conditions are favorable. Weather forecasters have become very good at identifying favorable conditions for the formation of tornados and monitoring their development. Forecasters issue watches and warnings when necessary; however, they are unable to provide accurate forecasts of the timing, strength, and exact location of tornados.

Winter Storms

Freezing rain and **sleet** occur when rain from relatively warm air falls through a layer of air that is below freezing (32°F/0°C). Sleet freezes solid before it hits the ground; however, freezing rain remains in a liquid state until it freezes on contact (Figure 6.10). Trees and utility infrastructure can suffer severe damage from freezing rain because their large surface areas allow the accumulation of many times their weight in ice (University of Illinois Atmospheric Sciences n.d.).

Table 6.2 The Enhanced Fujita Scale.

EF rating	Three-second gust	
	mph	km/h
0	65–85	105–137
1	86–110	138–177
2	111–135	178–214
3	136–165	215–266
4	166–200	267–322
5	200+	322+

Source: National Oceanic and Atmospheric Administration (www.spc.noaa.gov/faq/tornado/ef-scale.html)

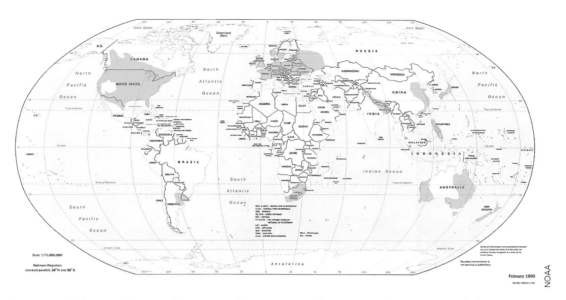

Figure 6.9 The shading on this map indicates areas where tornados are most likely to occur.

Figure 6.10 A utility arborist clears ice-laden branches from utility facilities.

Heavy, wet snows (generally defined as having less than a 10:1 snow-to-water equivalent) occur when temperatures are near freezing. In addition to being heavier than drier snows, these snows tend to cling to trees and infrastructure. Snow is likely to cause more damage when it occurs in spring and autumn, when leaves are present on deciduous trees. Coniferous trees in areas that receive frequent heavy snows are better adapted to snow loads than conifers in areas where snow is rare (Figure 6.11).

Figure 6.11 Conifers native to snowy areas are adapted to snow loads.

Local High-Wind Phenomena

Local terrain, climate, and seasons can combine to create unique local wind events. Terms such as "Chinook" (Rocky Mountains of U.S. and Canada), "Foehn" (Europe), "Santa Ana" (California and Mexico), "Scirocco" (Mediterranean), "Nor'easter" (eastern coast of U.S. and Canada), "Nor'wester" (New Zealand), and "Southerly Buster" (Australia) are just a sampling of names that refer to unique wind phenomena in various parts of the world. They often occur seasonally and come from a specific direction. An understanding of regional wind phenomena is part of ensuring successful selection, placement and management of trees in any region of the world, and in planning for storm response.

PRE-COORDINATION AND PREPARATION

Plans and preparations for storm response should be in place well before storms strike. While plans inevitably must be adjusted as conditions change, during a storm is not the time to make arrangements that could have been put in place ahead of time. Many preparations involve multiple parties from different companies and government agencies, all of whom will be extraordinarily busy during a storm response (Mullen 2013). Plans and contingencies should be developed and rehearsed both internally and with potential responders and suppliers. Scenarios should include various types of storms that are likely to occur and unexpected problems that must be solved.

Personnel identified in storm emergency plans should be prepared for deployment, either for a short-term, local response, or extended deployment to another region. Likewise, necessary equipment should be well-maintained and ready for long trips. Office and other personnel not deployed should also be prepared to adjust their routines as necessary, including being on call or extending their normal working hours to ensure that responding crews receive the support they need while they are away.

Chains of Command and the Incident Command System

To avoid misunderstandings, and to verify that any commitments made will be honored, each organization involved in a storm response should establish a clear chain of command. Alternates should be designated and their contact information should be made available in case primary personnel are unavailable. An unambiguous chain of command allows a responding organization to quickly adjust as the scale of the response changes or shifts focus, and to fit into an overall response structure, such as the **Incident Command System (ICS)** (Edison Electric Institute 2014).

The ICS is a well-established, standardized method for handling emergencies that is recognized internationally. Emergency responders in the public and private sectors have come to recognize that using an ICS assures better coordination with government emergency management agencies, such as Federal Emergency Management Agency (FEMA) in the United States, Public Safety Canada, and corresponding emergency management agencies in other countries. Use of an ICS can improve the overall effectiveness of emergency response by reducing friction or redundancy between participating organizations (Mullen 2013).

The U.S. Department of Transportation describes the ICS as a systematic method for "command, control, and coordination of emergency response" (United States Department of Transportation 2006). According to FEMA, the ICS allows multiple organizations, facilities, equipment, and personnel to operate in "a common organizational structure, designed to enable effective and efficient incident management." As providers of essential services, utilities are often included in ICS plans. By creating response plans and procedures that fit into the ICS structure, utilities, contractors, governments, and other organizations involved in a response can ensure better coordination of restoration efforts (Federal Emergency Management Agency Incident Command System Resource Center n.d.).

Emergency Operations Center

Most utilities, municipalities, and large tree contracting firms have designated storm processes and a central designated **emergency operations center (EOC)** where conditions are monitored and decisions are made (Figure 6.12). The size and scale of the EOC will vary depending on the size and type of organization, as well as the frequency and severity of storm events. Regardless of size, EOCs should have reliable communication, business continuity plans, backup power, links to weather and news services, and the capacity to accommodate necessary personnel during large-scale responses. Additionally, during major responses, on-site regional command centers may be required, often in the form of mobile offices, to further support response capabilities. Storm centers should be prepared to act in coordination with the ICS chain of command framework (Edison Electric Institute 2014).

Figure 6.12 An emergency operations center with communication links and backup systems.

Pre-Storm Communications Check

Organizations involved with storm response should ensure that contact information for all essential personnel is up-to-date. This information should be readily available internally and shared with other organizations likely to be involved with storm response. Key personnel from external organizations should be identified, including those in utilities, contractors, support services, suppliers, and first responders and government personnel at the federal, regional, and local levels.

Labor contracts may have clauses that require certain personnel to be called out first. Call-out lists should be kept updated and highlighted with this type of information.

Practice Drills

Many utilities and municipalities routinely perform practice **storm drills** so that the extended team of employees and suppliers understands the appropriate procedures, and to also identify opportunities to improve performance before storms strike. Drills should include various scenarios because real storms often impact nearby areas, affecting the availability of resources and the timing of the response (Electric Light and Power 2014).

Areas Vulnerable to Tropical Cyclones

This extraordinary composite map, consisting of actual storm tracks from 55 years of data, clearly illustrates the areas most likely to be affected by tropical cyclones in the Atlantic, Pacific, and Indian Ocean basins, in both the northern and southern hemispheres. Interestingly, most of South America is free of these storms due to cold waters west of the continent and typically unfavorable atmospheric conditions to the east.

In 1900, before such comprehensive historic data was available, Galveston was one of the largest cities in the state of Texas, U.S., with over 38,000 residents (Texas State Historical Association n.d.). Despite the fact that the city had been built on a barrier island with a maximum elevation of 8 feet (2.5 m), civic leaders had been led to believe that the waters of the Gulf of Mexico would not support a powerful hurricane. On the recommendation of the local Weather Bureau meteorologist, they opted not to build a seawall, and in fact removed protective sand dunes, using the material as fill. The 1900 Galveston hurricane, estimated at a Saffir-Simpson category 4, struck on September 8 with a storm surge that overwhelmed the island. No evacuation was ordered. Nearly all of the structures were destroyed, and an estimated 8,000 were killed (Larson 2000).

Tropical cyclones can maintain intensity as long as they are over warm water. After making landfall, they lose strength at a rate that depends on overall intensity, terrain, and forward speed (Australian Government Bureau of Meteorology n.d.). For example, in 2005, Hurricane Katrina made landfall and moved quickly inland over relatively flat terrain in the southern United States. Hurricane-force winds extended 200 miles (322 km) inland and tropical storm-force winds were felt 400 miles (644 km) from the coast. On the other hand, mountainous terrain, such as that found on the island of Taiwan, tends to degrade tropical cyclone intensity; however, that same terrain increases runoff from associated heavy rains and often contributes to the incidence of severe flooding (Yang et al.).

Damage to infrastructure is caused by both wind and flooding. Low-lying areas, such as barrier islands and river mouths, can be severely

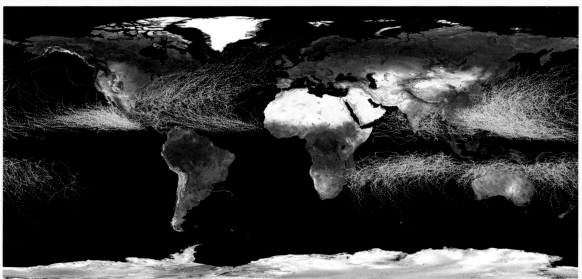

Composite showing all tropical cyclone tracks, 1950–2005.

affected by the combination of storm surge, heavy surf, and high winds. These areas are often heavily developed, resulting in costly damage. Wetlands are known to cushion the impact of tropical cyclones; however, many wetland areas have been damaged or removed by development, leaving some areas even more vulnerable.

Tropical cyclones have battered the world's coastal areas for millions of years—far longer than human beings and their assets have been around to be affected. Today, 40 percent of the world's population lives within 62 miles (100 km) of the coast (UN Atlas of the Oceans n.d.). Given this pattern, future strikes are inevitable, and with sea levels rising, their effects are likely to be compounded. With this understanding, mitigation strategies increasingly focus on the hardening of infrastructure and thorough preparation for swift response and recovery.

In planning for storm response, the ancillary effects of storms should be anticipated. These effects might be road closures due to flooding or emergency declarations, slippery and other adverse conditions, and the evacuation of residents from afflicted areas. Practice drills should account for these variables in planned response times (Figure 6.13).

Storm drills generally include a realistic scenario that involves the type, strength, and scale of the storm, and the extent of damage. Potential responders are contacted, estimates of available resources and response times are obtained, routes and **staging areas** for responding crews are designated, and other details are handled as they arise. In short, everything is addressed except the actual movement of personnel, supplies, and equipment (Next Era Energy Inc. 2013).

Potential responders should take these drills seriously and participate as if a real storm is occurring, including obtaining necessary releases of

Figure 6.13 Difficult traveling and working conditions should be anticipated when practicing responses.

crews and equipment from areas not affected by the hypothetical storm. Considerations for drills include:

- Accurate contact information internally and externally, including other utilities, contractors, and government officials
- Types and paths of storms (consider multiple storms, e.g., a second hurricane on a similar or dissimilar track)
- Seasonal considerations (appropriate clothing, lights, tire chains, or other necessary supplies)
- Estimated response times (accounting for extra distances required due to the incoming storm's effect on nearby areas)
- Routes taken by incoming crews (account for road restrictions, such as bridge load limits, height restrictions, weather conditions, and whether commercial traffic is permitted)
- Locations of staging areas
- Contingencies for disruption, including power outages, communications, displaced residents, and the availability of food, housing, fuel, and other support services
- Locations and capacities of regional hotel and motel facilities, as well as alternative facilities, including schools, community centers, or army barracks
- Providers of logistical services, such as climate-controlled tents, cots, sleeping trailers, catering, sanitation, and security
- Alternative communications—such as radios or satellite phones—in the event landline and cell phone networks are down

Identify Suppliers and Pre-Negotiate Terms

During a storm emergency, responses should not be hindered by confusion or disputes that could have been worked out ahead of time. Pre-negotiated agreements should be in place with suppliers, including billing terms. To reduce the likelihood of disputes and misunderstandings, some flexibility may be necessary to accommodate reasonable differences in the customary practices of responders.

Pay rates, equipment charges, and work rules may vary significantly between regions. Generally, it is understood that incoming personnel will receive the higher of either the worker's home pay rates or the established local rates, and that existing contract agreements of incoming workers will be honored. However, under certain emergency situations, and with the cooperation and consent of unions, some work rules may be suspended.

In some cases, especially in large or extended responses, storm work may require specialized support services or supplemental supplies, such as:

- Temporary housing (e.g., large tents, trailers)
- Mobile food, shower, and sanitation services
- Extra tools, spare parts, clothing, lights, or other materials
- Specialized equipment
- Fuel
- Security
- Entertainment

Suppliers of these items and services that are needed in a worst-case scenario should be identified and the terms of their delivery and payment put in place (Edison Electric Institute 2014).

Monitor Conditions and Pre-Mobilize

Storms like tornados or severe thunderstorms often strike quickly, while others such as tropical cyclones and ice storms may take several days to develop. Regardless, all personnel who could be involved with storm responses—including office support—should monitor weather conditions daily, especially during the times of year when storm response is most likely, and be prepared to respond as appropriate on short notice. Web-based service providers and weather alert services offer automatic notifications of weather events, as well as updates as changes occur.

As storms develop and storm watches and warnings are issued, management personnel should provide organizational momentum going into the response by verifying that pre-mobilization efforts are complete. Employees should be reminded that a response is likely. They should also be reminded of the professional demands typical of any storm response situation, both general (e.g., conduct while away) as well as specific (e.g., tire chains, warm clothing, extra food). Local coordinating personnel should stay in close contact with corresponding coordinators at utilities, government agencies, or contractors.

When it is evident that a storm will not strike a particular region but will instead affect nearby areas, and there are no additional threats, potential responders should remain at a high state of readiness, and managers should prepare to release an appropriate number of crews for a response if necessary.

Releases of Personnel and Equipment

Storm response is a cooperative endeavor that leverages the capabilities of multiple regions to more quickly restore local services in stricken areas. Utilities, municipalities, and other agencies requesting help recognize the value of assisting one another during emergencies because their management understands that limited resources will be available to assist when a storm hits them (Edison Electric Institute 2014).

Because most large-scale vegetation management work is contracted, requests for assistance are often directed to contractors. Typically, a request will include the number of personnel and the type of equipment needed for the expected workload, and a preferred time of arrival. For example, if there is extensive damage to electric transmission lines in wetlands, then specialized equipment and personnel suited for that type of work must be located and transported to the site. The amount and severity of damage caused by some large storms can create acute local labor and equipment shortages, which has necessitated travel distances in excess of 1,000 miles (1,600 km) or more for many responders (Edison Electric Institute 2014).

Requests for assistance to a responding organization should be directed to a designated coordinator or other contact person in the EOC. The coordinator should have ready access to all necessary contacts and other key information. Calls received after hours must be directed to that person or, in the event of a problem (e.g., dead phone, personal emergency), a designated alternate. The EOC systematically assembles a list of available personnel, equipment, and supervision with corresponding distances and estimated travel times. Travel time estimates should be realistic, accounting for mobilization time, lower truck speed limits, and other conditions such as weather or traffic that may slow progress (Edison Electric Institute 2014).

Many utilities are part of voluntary **mutual assistance programs** (Edison Electric Institute 2014). For example, part of the stated mission of The Southeast Electric Exchange is "coordination of storm restoration services to impacted member companies." As part of these agreements, contracted line-clearance crews may be assigned to other member utilities, along with line crews, as part of a mutual assistance effort.

To reduce the possibility of misunderstandings, utilities that participate in mutual assistance programs should be certain that their contractors are aware of any obligations that might apply to them under such arrangements. Contractors should be notified immediately when such agreements or obligations become effective and should be provided with timely notice when threats to member companies have passed, so that crews and equipment resources can be made available to utilities outside the mutual assistance program if needed.

It should be recognized that storm work performed by outside crews also represents work not done where the crews are based. As a courtesy to those who are assisting, those requesting assistance should ensure that they ask for only the resources necessary to restore service and that when restoration is complete, crews are released to return to their home utilities, or to assist elsewhere, unless other arrangements are made.

RESPONSE

Preparations and negotiations made ahead of time set the stage for an effective storm response. However, to assure that responses run smoothly,

Planning for the Unpredictable: The Uncertainty of the Track Forecast Cone

In August of 2011, Hurricane Irene swept up the East Coast of North America, providing a classic example of how the uncertainty of hurricane forecasting can affect storm response. The storm formed in the Caribbean, striking Puerto Rico on 21 August, and strengthening as it moved toward Miami, with the rest of Florida in the center of the National Hurricane Center's (NHC's) **track forecast cone**. Four days later, Irene was well east of Florida, barely within the earlier cone and with most of the east coast of North America in a new track forecast cone.

The NHC track forecast cone is based on the accuracy of previous forecasts and is intended to predict the track of the storm center with an accuracy rate of 67 percent (National Hurricane Center n.d.). This means that it is designed to be wrong one-third of the time, on average. To further complicate matters, the size of the cone is fixed, even though forecasters may be far more or less confident about their forecasts than the shape of the cone suggests. And of course, areas some distance from the storm center may be significantly affected, depending on the storm's strength and size. These vagaries have led some to dub the track forecast cone as "the uncertainty cone" (Satterfield 2013).

After a somewhat unpredictable start, Irene settled in on a course practically identical to the center of the 25 August track forecast cone. It brushed North Carolina and the Mid-Atlantic states and made another landfall on 28 August in New Jersey, New York, and Connecticut as a large tropical storm, before dumping torrential rains into northern New England and eastern Canada.

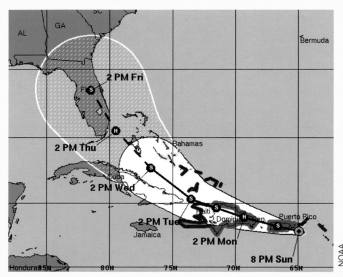
Hurricane Irene track forecast cone as of 21 August 2011.

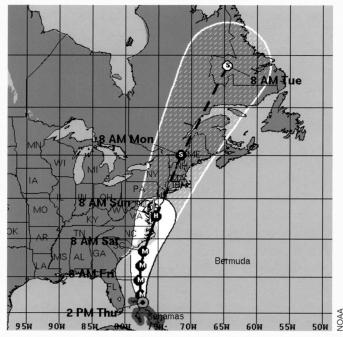
Hurricane Irene track forecast cone as of 25 August 2011.

At some point between 21 August and 30 August, Irene directly affected or threatened the entire eastern seaboard of North America. Irene's track, typical of storms in that region, effectively paralyzed potential responders from Florida, U.S. to Nova Scotia, Canada delaying the release of assistance personnel until the threat passed. To meet requests, crews were mobilized from areas as far as 700 miles (1,127 km) inland. Similar logistical difficulties sometimes occur with winter storms as they develop, strengthen, and track between regions.

As computer models and the science of forecasting improve, track forecast cones have narrowed. Still, when storms like Hurricane Irene threaten long stretches of populated coastlines, the inherent uncertainty of the track forecast cone will continue to vex those who plan for, and respond to, such storms.

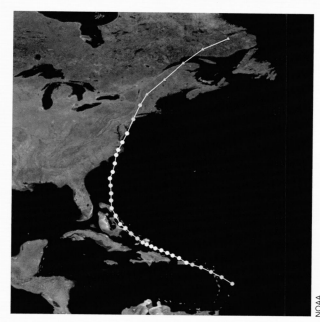

Actual track of Hurricane Irene, 20–30 August 2011.

services are efficiently restored, customers are satisfied, and every employee involved is treated well and returns home safely, the entire storm team must be focused and ready to adjust to a continuously changing environment.

The Crew

The crew is the basic work unit. Crews usually consist of two to five members (occasionally more) and corresponding equipment and tools, overseen by a foreperson. For storm purposes, crews are most commonly divided according to how they accomplish the work, i.e., with an aerial lift (often referred to as "bucket" crews) or by climbing ("manual"). Other specialty crews have additional capabilities based on their specific skills and the type of equipment used, such as four-wheel-drive or other off-road or mechanized capabilities.

From the perspective of a manager in an EOC, storm damage is viewed as a problem that is solved by responding with personnel and equipment (i.e., crews) that can get to the stricken areas, remove fallen trees and branches, and assist in restoring service in the most efficient time frame. Based on the extent of damage, the type of workforce required, and their ability to effectively organize the incoming workforce, the affected agency decides to request a certain number of crews and to incur the necessary costs. The responding company then assures that the necessary resources are marshaled and dispatched to their assigned destinations.

However, it is important for managers to remember that crews are made up of people, and that storm response has a big impact on the lives of the crewmembers and their families, especially when the response is to a distant location. When someone says, "Send the crews!" normal routines are suspended, and families are left behind—sometimes in the middle of the night, with little or no advance notice. Employees are suddenly working in unfamiliar territory under difficult conditions—perhaps hundreds of miles from home.

Under such circumstances and in what is often a chaotic environment, it is important to retain some sense of normalcy for employees. For this reason, it is customary practice to keep established crews together, and to move local supervision with crews. Not only does this help reduce stress for individual crew members, but it also helps maintain order and discipline.

Types of Storm Response

Very generally, storm responses can be divided into small- and large-scale categories. **Small-scale responses** are usually due to local outbreaks of severe weather and are handled by crews within their normal operating territory. Depending on the scale of damage, work might be completed within a few hours, or it could take several days. Employees may work extra hours, but they return home at night.

Large-scale responses are usually caused by large storm systems that overwhelm the capabilities of local crews, at which time "outside" crews are called in. These personnel must first travel to the storm and then be housed and fed for the length of their stay. Large-scale responses are inherently more complex, and often require coordination among multiple companies and with governmental agencies. They also generally trigger activation of utility mutual assistance agreements (Edison Electric Institute 2014).

Small-Scale Responses

A local severe storm may not make the national news but still may result in hardship for an affected community. While large-scale responses are more logistically challenging and receive more attention from management, the public, and the media, small-scale responses are more common and are cumulatively significant. It is important to have procedures in place to assure rapid small-scale response at any time of day or night.

Many small-scale responses simply entail directing one or more crews to mobilize and show up at designated locations within a certain time frame. Across a larger service territory, or with a larger storm, many crews may need to respond in multiple locations. In either case, management of both the contractor and the requesting agency must be in close communication, and the crews closest to the work crews must be located, mobilized, and provided with reporting destinations and job assignments when they arrive.

Most utilities and municipalities have clauses in their contracts that require their contractors to mobilize and have personnel on-site within a designated time after a call is received. Those times will vary depending on local circumstances, but are generally two hours or less for locally-based crews.

Large-Scale Responses

When large, powerful storm systems strike large metropolitan areas or heavily developed coastlines, responding personnel and equipment can number in the hundreds or thousands and may be required for extended periods (Figure 6.14). A safe and effective response requires the following elements:

- One or more sites to stage the operation
- Accommodations that ensure personnel receive adequate food and rest
- Roadworthy equipment with a full complement of tools
- Plenty of fuel
- Backup supplies of spare parts, PPE, clothing and other gear
- Reliable communications between all parties
- Personnel with specialized skills, such as experienced work planners and safety supervisors
- Experienced management to assure efficient coordination
- Professional employees, with a positive attitude and a commitment to work safely

Housing and Feeding Personnel

Locating all or some personnel in hotels may be an option; however, as the level of damage and number of incoming responders increases, the number of available rooms becomes limited. At such times, local residents who have no power or whose homes have been damaged may also be seeking accommodations. Additionally, hotel and

Figure 6.14 A large-scale response can be logistically challenging, and may require large staging areas.

meal services may not be available if power outages are widespread.

Assuring that basic needs are met following a catastrophic storm may require providing temporary accommodations (Figure 6.15). This usually requires the use of a number of specialty suppliers.

Alternatively, crews may be housed in schools, community centers, military barracks, or other large facilities. These can be provided with auxiliary power generators if necessary, and often have commercial kitchens that can be used to prepare meals. Outside suppliers may be needed to supply beds, catering, security, and other services. Ideally, contingencies for such arrangements are made in advance.

Pre-Staging Resources in Advance of Large Storms

Storms that develop gradually, such as hurricanes and ice storms, often provide enough time for areas in their projected path to respond in advance. As confidence about the path and intensity of a

Figure 6.15 Temporary accommodations, such as this sleeping trailer, may be necessary in large or extended storm responses.

A Town for Thousands Overnight

With many large-scale responses, the best way to handle the logistical challenges posed by the onslaught of a storm and thousands of responding personnel is to create temporary accommodations in one or more strategic locations. These facilities can be constructed quickly using a combination of temporary structures and trailers, and can include beds, food service, sanitation, showers, auxiliary power, laundry, entertainment, field offices, and security (Edison Electric Institute 2014).

While the notion of hundreds of people sleeping in large tents and trucks may not seem ideal, these types of accommodations need not be uncomfortable. In fact, a well-managed tent facility offers important advantages, including:

Accommodations for many personnel can be quickly erected as part of a large-scale storm response.

- **Oversight.** Centralization allows for ease of supervision, including start/stop times, equipment maintenance, distribution of supplies, and monitoring of crew behavior. It also provides a more secure environment.

- **Economies of scale.** Consolidation of services such as meals, lodging, and laundry in one location is usually more efficient than feeding and housing workers in hotels or in scattered remote locations.

- **Comfort.** Tents and trucks can be heated or air conditioned, even when utility services are not functioning in surrounding afflicted areas. Services such as laundry, sanitation and first-aid ensure that needs such as basic hygiene do not become problems.

- **Communication.** Safety bulletins, announcements, and other necessary communication can be easily provided to all employees in one location.

Disadvantages include:

- **Finding suitable locations.** Adequate space and access within a reasonable distance of the work may not be available.

- **Less flexibility.** As work locations change, time and fuel required to reach job sites may reduce efficiency.

- **Cost.** Initial setup cost makes these accommodations less useful for short-term or small-scale responses.

Once a site is identified, turn-key services can be provided by outside vendors, freeing up management personnel for other duties.

storm increases, managers must decide whether or not to **pre-stage** the storm to improve the speed and effectiveness of the restoration effort by strategically moving personnel and equipment before a storm strikes (Next Era Energy Inc. 2013; Edison Electric Institute 2014).

The decision to mobilize crews before a storm strikes must be carefully considered. Movement of crews is costly, delays planned work schedules, and disrupts the personal lives of employees. If storms do not live up to expectations, much time and effort are wasted. However, once there is high confidence in a strike, there may be a window of opportunity to position crews in advance of the storm. Considerations include:

- Expected severity of the storm (i.e., type of damage and number of customers likely affected)
- Reliability of weather forecasts
- Available internal resources
- Time required to assure safe transit, given the distance to be traveled and the path and intensity of the incoming storm
- Availability of useful staging areas out of the path of powerful storms

Sending Crews

When sending crews, whether for a day or an extended stay, it is essential that they arrive prepared to work, i.e., with a full complement of tools and supplies, and well-maintained, roadworthy equipment. It is the responsibility of sending management to ensure that crews know how to safely reach their assigned destination and are properly equipped for the conditions they will encounter at the storm site. The last thing a storm-stricken area needs is to scramble to find gear for ill-equipped crews (see *What to Bring to a Storm?* on p. 185).

Rosters. Complete, accurate rosters of personnel and the equipment to which they are assigned must be provided to the destination utility or agency. Such information is important for many reasons, including:

- Safety considerations
- Verifying that employees are where they are supposed to be
- Assuring accuracy of accounting and billing
- Contacting employees for emergency or personal reasons

Routing and Navigation. Routes taken by crews should be planned, taking into account any road restrictions and traffic problems caused by the storm. When going to an area that is being evacuated, it is best to check with authorities to ensure that planned routes are open and passable.

An important consideration is bridge weight restrictions and clearances, which do not typically appear on consumer-oriented maps and navigation aids. Crews operating in unfamiliar areas should be provided with information about such restrictions, whether on paper maps or in the form of a commercial navigation device or application.

The location of most fleet vehicles is tracked with GPS systems, which aid navigation and ensure that crews can be quickly located for any reason (Edison Electric Institute 2014). With these systems in place and working, it should not be necessary to convoy. Location information should be used to ensure that crews and equipment are in expected locations and are following approved routes (Figure 6.16). Access to this information is routinely provided to receiving utilities and agencies.

Unusual Conditions. During large-scale responses crews are sometimes assigned to work in unfamiliar environments, especially when personnel are moved long distances. For example, crews from a warm location may respond to a winter storm, where warm clothing would be essential. Likewise, crews from cooler climates may encounter unfamiliar trees such as palms. It is important to provide crews with necessary clothing, supplies, and information that may be needed in unfamiliar work environments, and to ensure that safety information about these situations is provided as necessary in stand-downs, tailgates, or job briefings (additional safety considerations appear below).

Brush Disposal. Most utilities instruct vegetation management personnel to leave felled trees and branches on site during storm responses; in fact, brush chippers are typically not brought on storm work (Figure 6.17). Customers who are

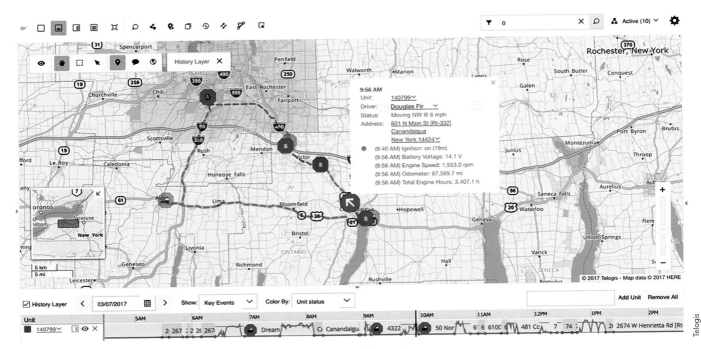

Figure 6.16 GPS tracking ensures that crews are in expected locations.

accustomed to typical utility pruning operations, where cut brush is chipped and removed from the site, may be quite surprised to find that brush disposal following a storm is their responsibility. Complaints regarding this policy can be reduced by ensuring that the reasons are made clear to the public and to media outlets. Most customers will be more accepting if it is understood that (1) the trees and branches belong to the property owners, and (2) the effort required to chip and haul brush is better spent on restoration (Edison Electric Institute 2014).

Communication. While extended cellular telephone outages are rare, it is not uncommon for normal cell phone coverage, as well as land lines, to be interrupted or unreliable in certain areas following large storms. Radios or satellite phones may be required in certain circumstances (Edison Electric Institute 2014).

Receiving Crews

While it is necessary to secure personnel and equipment to restore services rapidly, it is critically important for requesting agencies to ensure that incoming crews are properly accommodated when

Figure 6.17 During storm response, debris is usually left on-site to allow crews to focus entirely on restoration of services.

they arrive. It is very disheartening for personnel to spend long hours on the road, only to find a chaotic environment with no place to eat and rest when they arrive at the storm. In fact, without adequate

What to Bring to a Storm?

Anyone sending crews should ensure that crews are ready to work when they arrive. Following is a checklist of items to include or consider when sending crews to a storm:

- Potable water
- Extra food and snacks (non-perishable, pre-packaged)
- Extra tools and spare parts (chain saws, handsaws, pruners, files, chains, bars, ropes, saddles, straps, carabiners, etc.)
- PPE, including hardhats, safety glasses, ear protection, gloves, vests, etc.)
- Rain gear
- Extra fuel and bar oil for saws
- Signs and cones
- Seasonal items (insect repellent, tire chains, warm outerwear, etc.)
- Lights for night work
- Fully stocked first-aid kits
- Flashlights and batteries
- Navigation system and paper maps
- Cell phones and chargers (consider radio system as backup)
- Purchasing cards for fuel and miscellaneous supplies

Employees are responsible for their own personal items, such as:

- Several changes of clothing, appropriate for conditions
- Footwear, including work boots and comfortable after-work shoes
- Sunscreen
- Medications
- Glasses (and spares)
- Personal hygiene items
- Personal cell phone
- Headphones for music and podcasts
- Reading material

Some things are *not* appropriate in the environments that will be encountered during storm responses. Items that are dangerous or could elicit bad behavior should be left behind, such as:

- Firearms or weapons of any kind
- Alcohol or illegal drugs
- Valuables (jewelry, expensive watches, quantities of cash, or other items likely to be stolen)

---

food and rest, it is not safe for employees to work, potentially making them more of a liability than an asset.

Before final requests are made for outside assistance, requesting agencies should ensure that all incoming personnel will be provided with meals and a safe and reasonably comfortable place to sleep (Edison Electric Institute 2014).

It is also important to ensure that incoming personnel are quickly oriented and assigned work. Not only does this ensure that services are restored as quickly as possible, but it also reduces the chances that idle personnel will be observed by the public or media outlets, resulting in negative publicity.

Emergency Declarations

What exactly constitutes a state of emergency varies by country and local jurisdiction; however, most national and local governments can officially declare a state of emergency when extraordinary conditions—such as those found in storms—disrupt normal activities and cause damage to infrastructure (Federal Emergency Management

Agency n.d.). At such times, additional support personnel, loans, or other funding may become available, and certain laws or regulations may be suspended, including:

- Licensing or permit requirements for arborists
- Certain commercial driver requirements (e.g., limits on hours of service)
- Collection of taxes and fees on responding trucks and other equipment
- Traffic flow, especially in evacuation areas
- International border crossing protocols, including inspections and declarations

Organizations such as electric utilities may declare emergencies independent of official government declarations. During utility declarations, and during local government-declared emergencies, it should be assumed that wider regional or national regulations will still be in effect. Responders should make no assumptions, and should be aware of which rules are in place before departing.

After destructive storms, debris removal and disposal work may be eligible for funding as part of government emergency or **disaster declarations**. Companies with these capabilities may have an opportunity to participate in these efforts. In some cases, especially when government officials issue disaster declarations, external funding may become available for brush disposal (Edison Electric Institute 2014).

Special Safety Concerns

For employees in a storm response, all of life's normal routines have been turned upside down. Working long hours far from home in unfamiliar environments and eating and sleeping in unusual patterns can lead to unsafe working conditions unless proactive measures are taken.

If a concern is identified, perhaps due to an incident or a near-miss, and immediate dissemination of this information could prevent further incidents, a safety "stand-down" may be called, in which all operations are brought to a stop, and the relevant information is immediately provided to everyone. Otherwise, employees should be reminded of the specific conditions likely to be encountered in a storm response, and supervision should pay special attention to safety. Job briefings should include any unique recognizable hazards specific to the circumstances.

The following are brief descriptions of some unique aspects of storm response that have the potential to affect safety awareness and performance.

Fatigue. Lack of sleep, nutrition, inadequate hydration, and working long hours, cumulatively, can reduce an employee's situational awareness. It is important to ensure that employees have ample water, food, and rest. Watch for signs of fatigue and intervene by pulling employees off the job. In some extended responses it may be necessary to rotate in fresh workers.

Working away for extended time. Once the initial excitement of being involved in a response wears off, some employees may become distracted by homesickness or problems at home. Supervision should maintain a dialog with employees and recognize when their effectiveness is being compromised by the stress of being away. Again, employees can be rotated out as needed.

Working in difficult or unusual conditions. Following storms, poor weather may persist, or extreme changes may occur. Tropical systems are often associated with hot, muggy weather, and likewise, winter systems may bring persistent cold, accompanied by additional snow or ice. Each of these conditions presents its own set of hazards to which all employees must adapt, and to which not all employees are accustomed.

Electrical hazard communication protocol. Even when power is out, lines can be energized by back feed from improperly connected home generators. For this reason alone, workers must treat all lines as if they were energized. In addition, utilities have varying protocols for grounding and isolating sections of line that are being repaired. All personnel should be made aware of local procedures and who to contact if there is any question about if or when a line may be energized.

Animals and insects. After a major storm, humans aren't the only individuals displaced. Pets may be lost or left behind and should be approached with caution. Wildlife may also have fled flooded or damaged areas and may be roaming in unusual or unfamiliar areas. This includes everything from large animals, like bears and alligators, to snakes and even insects, like bees and hornets (Figure 6.18). Employees should be reminded that these situations can be dangerous, and that wildlife should be handled only by licensed professionals.

Different kinds of trees and other vegetation. Responding personnel may not be familiar with the trees in the local response area. Wood strength, branching patterns, the presence of thorns and other characteristics may present unknown hazards to visiting crews. Likewise, noxious plants, such as poison ivy or giant hogweed, pose greater risk to employees who do not encounter these plants in their home regions.

Wood under tension and other hazards. Fallen trees and branches may be entangled, and partially fallen or leaning trees or hanging branches may be common. In such conditions, employees must be especially aware of the possibility of unexpected movement when loads are released by cuts (Figure 6.19).

Figure 6.18 Wild animals, such as this alligator, may be displaced and found in unexpected locations following storms.

Figure 6.19 Wood under tension should be handled with extreme caution.

Driving. It is one thing to drive a truck a few miles from the usual parking location to a local job site, but it is another situation altogether to drive that same truck hundreds of miles on unfamiliar roads and in what may be poor or deteriorating conditions. Whether traveling to or from a storm, drivers should maintain awareness and should seek assistance from fellow employees when needed. Rest stops should be scheduled and drivers rotated as allowable.

These are just a few examples of common safety considerations. Employees should always be aware of their surroundings, including the actions of their fellow employees, and take whatever steps are necessary to prevent incidents.

Saving Damaged Trees

Trees sustain a range of damage following strong storms, from complete failure to minor breakage of the smallest branches. Many trees sustain severe damage but remain standing. The first impression may be that such trees are damaged beyond repair, and in some cases this may be true. However, severely damaged trees can often be restored to an acceptable form in far less time than what is required to plant and grow comparable trees of the same size. Considering the relative value added to communities by large trees, it is worthwhile to retain trees that have the potential to be restored.

During initial storm response, utility arborists are not expected to devote resources to **restoration pruning**. However, they should make reasonable efforts to avoid unnecessary damage to trees that could be retained. Responding arborists should make a preliminary assessment of standing trees, then prune or remove as necessary to allow access and restore critical services, understanding that trees are a valuable resource as communities recover from **disasters**. Subsequently, more comprehensive assessments can be performed to determine suitability for retention and to specify restoration measures as needed.

Documentation

For accounting purposes, storm work and routine work must be delineated. Most utilities and regulatory agencies have definitions and requirements that make the matter clear in their jurisdiction; however, there is no broad industry standard that defines what constitutes a storm emergency. In most areas, an isolated outage caused by a branch failure on a breezy day is generally not recorded as a storm emergency; however, a severe thunderstorm that fells trees, causes widespread outages and requires a significant departure from normal operations is usually reported as a storm emergency.

Storm response efforts may be complex and often seem chaotic, but this is all the more reason to be sure that accurate records are maintained. The response should be carefully documented from the start, with information shared between parties as appropriate, including personnel involved, equipment used, routes traveled, work assignments and locations, and time and all associated expenses. Keeping track of this information ensures that the location and activities of employees are known and can be accounted for when necessary (Edison Electric Institute 2014).

At some point, responding companies will ask for payment for work performed in storm response and may be reimbursed—depending on contract terms—for all or some expenses incurred along the way. Even for a few personnel, keeping an accurate accounting of hours worked, miles traveled, and costs for food, fuel, lodging, and other expenses can be daunting; however, it is essential that this is accomplished regardless of the size of the response. If not, disputes are more common, some legitimate expenses may not be reimbursed, and additional expenses will be incurred in later audits. More importantly, a loss of trust between parties may result.

To avoid confusion, a pre-negotiation of terms, including agreements on what expenses are reimbursable, will help streamline reporting. Contracts should reflect any such agreements that are in place, or absent those, should clearly state terms, including which expenses are included. The use of dedicated corporate credit cards can improve documentation and accounting, as can pre-negotiated turn-key contracts with suppliers of services and materials.

Media Relations During Storm Responses

Emergency declarations and the associated flurry of activities may attract intense public and media scrutiny, especially after large storms and during lengthy power outages. At such times it is not uncommon for members of the media to approach field personnel and ask questions. Unfortunately, many answers—even those provided with the best of intentions—can be misunderstood or deliberately distorted by the questioner. Just one stray comment from a tired, disgruntled, or homesick worker can result in negative publicity that reflects poorly on the entire response effort. At the same time, answering with "no comment" is often perceived as an attempt to hide something. For this reason, it is good practice to provide basic, pre-approved statements for employees (for example: "We are all working hard to get things back to normal" or "My assignment is to clear this street and we expect to be finished soon but I don't know anything about the rest of the work"). More substantive statements must first be vetted and provided only by designated personnel.

Also, most employees are able to instantly post photographs, video, and commentary on various social media outlets. While usually done without bad intent, such postings may contain information that is confidential or proprietary, or that can be taken out of context. It should be made clear to employees that they should not post photographs, video, or descriptions of storm damage or response efforts in any public forum, including social media outlets, without prior approval.

Members of the Public

During an emergency with widespread damage and power outages, property owners are more likely to be at home cleaning up, making repairs, or simply protecting their belongings. When restoration teams show up, it is natural for homeowners and neighbors to ask questions about the response, such as when power will be restored, or to ask crews for favors, such as removing damaged trees on private property.

In these situations, employees have no way of knowing who the property owner is or what their affiliations may be. Friendly small talk is fine, as is providing answers to basic questions about the work on particular properties, but arborists under contract should make no commitments other than the completion of their assigned work and should not comment on behalf of anyone, including utilities, government agencies, or contractors—including statements about how the overall response is going, when power will be restored, or the extent of the damage. As with reporters, such questions should be answered only with approved answers, or referred to designated personnel. Often, during storm response, an "I don't know" answer is best—it cannot be misunderstood and discourages further questioning.

Winding Down a Response

Releasing Crews

As the length of a storm response increases, workloads eventually decline, and requesting agencies begin to release storm personnel. However, more severely affected areas may still need assistance. Decisions about reassigning crews should take into account the readiness and ability of employees to accept additional assignments, as well as whether their service is required at home.

If crews have many miles to travel, it is customary for hosting agencies to provide them with a night's rest prior to releasing them, so that crews can avoid beginning the journey home right after working a long shift.

Arriving Home Safely

Long drives back to home territories must be approached with the same caution as with any other trip. In addition, some of the regulatory suspensions that were part of initial emergency declarations may no longer be in effect. Employees, looking forward to going home, may be tempted to push the limits of both speed and fatigue, especially after extended responses. Personnel should be advised to take their time and comply with all rules and regulations to arrive home safely.

INCORPORATING LESSONS LEARNED

Storm response pushes operations to the maximum, challenging utility arborists and the companies involved at every level. Whenever any system is performing at top capacity—machines, athletes, organizations—weaknesses are more evident. Unexpected events occur, and inevitably, mistakes are made, sometimes with costly consequences.

After every storm, the organizations involved should take time to analyze the response effort and determine the causes of both successes and failures. These findings can then be incorporated into preparations for the next response. This takes special effort, as the natural inclination is to focus on picking up normal operations where they were left off. However, considering the importance of storm work, the value of incorporating lessons learned into future responses is too great to ignore (Federal Emergency Management Agency 2013).

Managers need not wait until the storm response is over before incorporating lessons learned, especially regarding safety. When necessary, safety stand-downs (as mentioned above) are a way of immediately incorporating important new information into operations.

RECOGNITION OF EMPLOYEES

A successful storm response is driven by the efforts of employees at all levels. Evaluations and debriefings are a good time to consider the accomplishments of the employees who participated in the response and to devise appropriate recognition. Recognition often takes the form of commemorative items such as hats or t-shirts, as well as write-ups in company newsletters and award presentations at company meetings.

SUMMARY

Where trees and utility facilities exist in proximity, there is risk of service outages and damage from tree failure during storms. As the accuracy of storm forecasting and knowledge of tree failure patterns improve, the ability to assess and mitigate the risk posed by storms increases. Understanding the characteristics of storms and how they affect the likelihood of tree failure and their potential impact on utility infrastructure allows for improved mitigation and better preparedness for storm response.

The unusual conditions inherent in storm responses can push workers and their support functions to the extreme. Ignoring details because of high workload or disorganization inevitably results in greater problems in the long run. Taking the time to rehearse procedures with employees and clients and ensuring that employees, equipment, and tools are ready to be deployed are all critical steps in planning and preparing for a safe and effective storm response. In larger events, the team may include thousands of people and multiple entities, both public and private, working under a unified structure.

After major storms, the specialized skills provided by utility arborists are in short supply. For this reason, storm response should be regarded as an essential part of every utility arborist's job, whether responding at the site of the storm or coordinating from an office. In all cases, difficult conditions, unexpected challenges, and long hours should be expected.

Storm response represents the collective hard work of many individuals. Responses are most successful when all parties work together, focused on safely completing the required tasks, and constantly adapting to changing conditions. In this way, essential services are restored, and all employees return home safely.

CHAPTER 6 WORKBOOK

Fill in the Blank

1. _____ freezes solid before it hits the ground; _____ _____ remains in a liquid state until it freezes on contact.

2. The _____ _____ _____ is a systematic method for command, control, and coordination of emergency response.

3. Most utilities, municipalities, and large tree contracting firms have designated storm processes and a designated _____ _____ _____ where conditions are monitored and decisions are made.

4. Personnel from external organizations who travel to a storm area are generally paid the higher of either their _____ ____ rates or the established _____ ____ rates.

5. In the United States, the National Hurricane Center track forecast cone is intended to predict the track of a storm with an accuracy rate of ___ percent.

6. Removal of _____ following a storm is typically the responsibility of the customer.

7. A safety _____-_____ is a proactive measure in which all operations are brought to a halt until relevant safety information can be provided to response workers.

8. Organizations such as electric utilities may declare an _____ independent of official government declarations.

Multiple Choice

1. Enhanced Fujita ratings are based on
   a. estimated damage before the storm
   b. wind speeds during the storm
   c. observations during the storm
   d. observed damage following the storm

2. Rain and snow on dense foliage can increase the likelihood of tree failure.
   a. True
   b. False

3. Heavy, wet snows have a snow-to-water equivalent that is
   a. 10:1 or less
   b. 15:1
   c. 20:1
   d. 30:1 or more

4. On the Saffir-Simpson Hurricane Wind Scale, a storm must be at least which category to be considered major?
   a. 5
   b. 4
   c. 3
   d. 2

5. A standardized method for improving the overall effectiveness of emergency response is known as the
   a. incident coordination system
   b. incident command system
   c. emergency command system
   d. disaster response system

6. Which of the following is *not* an advantage of providing temporary accommodations for a large-scale disaster response?
   a. oversight
   b. economies of scale
   c. communication
   d. setup cost

7. Coniferous trees that receive frequent, heavy snows are better adapted to snow loading than such trees in areas where snow is rare.
   a. True
   b. False

CHALLENGE QUESTIONS

1. Outline the considerations for practice drills and how they might apply to storms in your area.

2. Why is the track forecast cone is sometimes called the uncertainty cone?

3. Explain the differences between emergency declarations made by state or local governments and those made by utilities.

4. Describe special safety concerns unique to storm response and some recommended ways to ensure safety on the storm job sites.

5. Discuss why it is critical to maintain accurate documentation for all activities that take place before, during, and after a storm response.

7
Communications

OBJECTIVES

- Explain how vegetation management activities can influence customers' perception of the utility brand.
- Recognize different stakeholder interests in utility vegetation management.
- Differentiate the roles of public relations, customer relations, and customer service.
- Describe various communication methods for interacting with the public.
- Use effective communication techniques when talking with customers to ease their concerns.

KEY TERMS

- active listening
- brand
- certification
- credential
- customer relations
- customer service
- desensitization
- empathy
- friendly language
- news media
- outreach
- press release
- public relations
- qualification
- social media
- stakeholders
- targeted messaging

INTRODUCTION

Utility vegetation management (VM) practices are often a source of controversy (Electric Light and Power 2006). While specific reasons vary, many concerns can be traced to a lack of basic communication between various stakeholders. Providing information about the benefits of VM, how it will be performed, and addressing specific customer questions and concerns improves understanding of the value of VM (Kuhns and Reiter 2007). When utility arborists practice effective communications, it is possible to avoid controversies, generate greater support for the program both internally and externally, reduce the number of complaints and associated costs, and improve the overall image of the utility.

Establishing good communications with stakeholders—such as residential customers, businesses, and communities served by utilities—is essential. Most stakeholders are not fully aware of the importance of utility arboriculture in providing safe, reliable electric service (Figure 7.1). This includes not only ratepayers, but often other departments within utilities and the government regulatory agencies that oversee utilities, as well.

Effective customer communications involves personnel at many levels in both utility and

Figure 7.1 Trees are ubiquitous in urban environments. Failures and impacts to overhead utility targets are inevitable; however, many stakeholders are unaware of the potential impact of trees on utility service.

contractor operations. How well they interact and craft and deliver relevant information can affect the success of other communications efforts. Furthermore, for many utility arborists, frequent public contact and high visibility are the norm, and once completed, their work is always on display (Figure 7.2). Ensuring that customer concerns are minimized, and that questions and requests are answered and acknowledged, requires training for both utility and contracted personnel, as well as the use of other verbal and written communications tools, including media relations and social media.

IMPORTANCE OF CUSTOMER COMMUNICATIONS

Frequent customer contact provides a typical VM employee, whether employed by a utility or a contractor, with the potential for thousands of public interactions over the course of a year. Across a large operation, this can add up to millions of customer impressions as employees drive, control traffic, knock on doors, return phone calls, answer questions, and perform work on customer properties. With some basic training, the vast majority of these impressions can be positive (Rawson 2013). Other utility departments, especially public relations and marketing, should be aware of both the value of VM in enhancing the quality of utility service, and the influence of VM personnel in shaping the public image of the utility.

Monitoring Performance

In industries where consumers can choose from multiple service providers, companies earn customer loyalty by providing high-quality service and competitive pricing (Ball 2004). However, in most areas, electric utility customers have little or no choice about who provides the services and how maintenance services are scheduled and performed.

Figure 7.2 Utility arborists should remember that their work is always on display for all to see.

In the absence of competition, government regulatory agencies oversee utility performance, including even the need for timely and effective communication (Commonwealth of Massachusetts n.d.). To gauge their performance, utilities may survey their customers. For more objective comparisons, there are reports available from companies that monitor consumer opinions of utility company performance (J.D. Power 2014).

Customer Perception

Research has demonstrated what most experienced utility arborists have known for a long time: utility customers often do not fully understand the need for VM and frequently do not like the appearance of the trees or the rights-of-way after the work is completed (Kuhns and Reiter 2007). It is also well established that customers place a high value on service reliability (Sullivan 2011). At the same time, data increasingly shows that VM is critical in protecting the integrity of the electric grid, from bulk transmission lines to individual service drops (Edison Electric Institute 2014). Finally, most utilities list public safety as a reason for maintaining vegetation (though customers often underestimate the risk) (Figure 7.3). With such a disconnect between the actual importance of the service and customer perception of its value, it is easier to understand why VM activities are often underappreciated, and why improved communication of the facts is essential.

Customer Dissatisfaction

In general, customers who have a negative experience for any reason are more likely to complain (RightNow Technologies 2010), though many do not complain directly to the utility. Contractors often receive and handle complaints without utility involvement. Many customers simply complain to neighbors, friends, and family—who are frequently also utility customers. The use of social media can greatly accelerate the rate at which information is transmitted, whether accurate or not. In this way, the level of customer dissatisfaction may be underestimated by the utility.

Widespread dissatisfaction can lead to lower expectations about the utility's performance, which in turn results in a higher rate of complaints. Handling complaints, whether by contractor or utility, takes valuable management time

Figure 7.3 Members of the public are often not fully aware of the risk posed by trees growing near utility lines until contact occurs.

away from other critical functions (Gore 1996). But higher complaint rates are really just a symptom of a much larger problem. Ultimately, high levels of customer dissatisfaction—whether justified or not—result in reduced public trust and confidence, and damage to the utility **brand**.

In addition to permeating and gaining traction on social media, high rates of customer dissatisfaction and complaints may attract the attention of news media outlets. The resulting negative publicity further amplifies damage to the brand of the utility, and has resulted in regulatory agencies placing restrictions or additional requirements on utility VM activities (Indiana Regulatory Commission 2012).

To avoid the downward spiral that leads to reduced customer satisfaction, negative publicity, brand damage, regulatory interference, and the associated higher costs, utilities and their contractors should identify the reasons for customer dissatisfaction, then develop strategies to better inform customers about the benefits of VM activities (Kuhns and Reiter 2007). One successful strategy is to provide employees who have frequent customer contact with training in basic communication skills.

IDENTIFYING STAKEHOLDER INTERESTS

By their very nature, electric utilities have many **stakeholders**. Those in charge of VM programs should develop good working relationships with various stakeholders, including residential and commercial customers, communities, property owners, media outlets, and public officials, as well as the stakeholders within utility companies (Miller 2014). Establishing good relationships with all stakeholders requires coordination of utility efforts across departments, including contractors and their employees. This should incorporate development of **targeted messaging** for different stakeholder groups.

Internal Stakeholders

Internal stakeholders include utility and contractor personnel who manage and perform VM services, such as utility pruning, integrated vegetation management (IVM), work planning, and auditing. Other departments within utilities, some of which may compete with VM for funding, and those with an influence over VM funding and contract terms, such as executive, finance, and purchasing, are also involved. Finally, there are the utility owners, who may be stockholders (in the case of investor-owned utilities), customers (as with cooperative systems), or governments (as with municipally-owned, regional or federal systems).

Utilities and Employees

Stakeholders at utility companies include any person or department influenced by VM activities, such as safety, service reliability, power quality, maintenance, and public relations. Unfortunately, the value added by a well-run VM program is often not well understood by other stakeholders within the utility. This phenomenon is illustrated by the story of the utility CEO who asked the line maintenance supervisor why so much money was being spent on line clearance when so few outages were caused by trees. The answer (of course) was that trees were not a problem *because* the VM program was well funded and highly effective.

Ensuring that all stakeholders within utility companies are aware of the benefits of proper VM practices will generate greater internal support for VM programs, help to secure funding, and encourage the hiring of trained professionals to both manage and carry out the work.

Contractors and Employees

Contract employees comprise the majority of the utility VM workforce. Front-line contract employees report that customers often berate their work and complain to them about the appearance of trees. At such times, the natural inclination is to become defensive and explain, "This is what the utility wants," or "We're just trying to do our jobs." Many of these employees are unaware of the actual value of the work they do and how this might be communicated to customers.

The high rate of contracting can exacerbate what is often already an inherent cultural separation between VM programs and utility management. Some utilities tend to view VM budgets as hedges against revenue fluctuations (Perry 1977).

Indeed, large VM budgets make tempting targets when overall budgets are tight, and the effects of cuts are directly felt by relatively few utility employees. However, the effect of fluctuating budgets on contract employees is significant and costly in terms of nearly all measures of performance, including safety, work quality, and customer satisfaction (Grayson n.d.).

High turnover rates in the VM workforce place a large training burden on contractors. Safety training and other basic job skills are given a higher priority than communications training. However, low job satisfaction is known to increase employee turnover (Liu et al. 2012). Furthermore, work that employees believe to be meaningful affects job satisfaction and employee well-being, and attracts more desirable employees (Steger 2012). With customers regularly decrying their work, and utility management culturally separated, many employees feel that their efforts are not appreciated, or are unaware that their work has any significant value at all.

Contract design can play a role in the integration of VM programs in utility culture (Neal n.d.). Contracts that create stable levels of employment and reward performance in customer satisfaction encourage employees to improve their customer communication skills. Managers at utilities and contracting companies should recognize this and be able to quantify the value of an experienced professional workforce (Johns 1981).

Communications training in VM should emphasize the value of the work performed. Employees should also be encouraged to pursue professional development—through appropriate training, **certifications**, **qualifications**, and licensing—and be offered recognition or incentives for achievement of these professional milestones. Contractors should work with their counterparts at utilities to create a business case for the VM program, and obtain a commitment for long-term support (Johns 1981; Neal n.d.).

Utility Owners and Investors

Utilities can be publicly traded, privately held, owned cooperatively, or owned by governments. Each type of ownership has its own set of communications challenges. For example, in the case of large investor-owned utilities (IOUs), the stockholders' primary interest is the financial performance of the company, usually measured in terms of profits and stockholder equity, as opposed to cooperatives, which are not-for-profit, and government-owned utilities, where customers have a direct interest in both financial and customer service performance.

Targeted messaging regarding VM will be quite different for these diverse owners. IOUs are usually subject to intensive regulatory oversight, which keeps investor-owners a step removed from operational details like vegetation management budgets. On the other hand, customers of cooperatively- and government-owned utilities have a direct influence on utility governance, so concerns about vegetation management practices can become significant political concerns for managers (Martindale 2015). In any case, utility managers who understand the value of VM can make more informed decisions about what is in the best interest of utility owners (Deric and Hollenbaugh 2003).

External Stakeholders

External stakeholders include residential and commercial utility customers, along with landowners with property adjacent to utility facilities. But many other interests are also stakeholders, including local communities, media outlets that may cover stories involving utilities, suppliers of equipment, materials and services used in the VM industry, and others that could be affected by utility VM practices. External stakeholders also include regulating agencies at federal and regional levels, which can have an enormous influence on how much funding utilities budget for VM. The development of targeted messaging efforts should be considered and prioritized for various external stakeholders as shown in Table 7.1. Stakeholder interests often overlap; therefore the information provided in targeted messaging should be customized based both on the work performed and the interests of various stakeholders. In every case, the stakeholder should be provided with a contact to obtain more information. Records of information provided to stakeholder groups should be maintained.

Table 7.1 Examples of targeted messaging for external stakeholders.

Stakeholder	Information provided	Suggested formats
Customer or landowner	Basic information about the work; why and when it will be performed	Brochure, email, text message, site visit, social media
City arborist	Neighborhoods affected, project overview, work specification	Email, meeting
Local government or community group	Project overview, purpose, schedule	Public meeting, brochure, email, social media
Environmental organization	Purpose of work, environmental benefits, opportunities for partnerships	Email, meeting, site visit
Media outlet	Project overview, benefits, sample script	Press release, email, social media, public meeting
Government regulator	Project overview, costs, benefits to ratepayers	Letter, email, meeting

Residential Customers

Establishing a good relationship with residential customers may seem like a long shot, especially considering that most customers have not asked for the service, do not understand why it is being done, and dislike the resulting appearance of their trees. However, communication can be improved with targeted messaging that emphasizes benefits to the customer, and a courteous, professional workforce. This ensures that a greater number of customers at least understand and possibly appreciate the service (Kuhns and Reiter 2007).

Commercial Customers

Most businesses cannot function without reliable power; indeed, surveys have shown that retail electric customers are willing to pay up to ten times the retail price of electricity to avoid a service interruption (Momoh 2010). During an outage, sales or productivity often stop completely, while fixed costs, such as rents, mortgage, depreciation, taxes, and payroll continue unabated. Therefore, businesses have an interest in supporting reliability improvement initiatives like VM. However, it is important for vegetation managers to present a business case that quantifies and explains the benefits of VM activities (J.W. Goodfellow, personal communication).

Communities

Local communities often struggle to balance their interest in a reliable power supply with the cumulative benefits provided by urban forests (Figure 7.4). While it may seem as if there is little common ground, there are common interests, including:

- Maintaining existing trees to minimize risk to all urban infrastructure
- Maximizing environmental benefits, such as reduced urban heat islands and airborne particulates, and increased carbon sequestration
- Creating attractive communities with a good quality of life

Environmental Organizations

Environmental organizations are often critical of utility VM efforts. Concerns about the appearance of trees, tree removal, the use of herbicides and the effect on wildlife are often expressed. While some potential conflicts may not be completely resolved, in many cases there is opportunity to demonstrate good stewardship, and to collaborate in achieving common goals (Figure 7.5). Utility arborists should reach out to concerned groups to identify and mitigate their concerns as much as possible. In so doing they can often win the acceptance—

Figure 7.4 A well-managed, healthy urban forest is beneficial to both the local community and the utility.

and even the good will—of these groups and their members.

Landowners

Owners of land adjacent to utility facilities that are affected by right-of-way (ROW) construction and maintenance may object to these activities on or near their property. This is especially true along transmission corridors, where landowners are not always customers of the utility (Furby 1987). These interests are often well organized and vocal in their opposition, with newsletters and social media campaigns, posing major communications challenges for utilities (Knowles and Branley 2017).

Proactive communication and outreach by utilities can reduce the negative impact of such stakeholder opposition. For example, in many cases, the environmental benefits of responsible ROW management have not been adequately explained. Demonstration projects and partnerships with environmental organizations can improve acceptance by skeptical landowners (Figure 7.5).

Media Outlets

News media, including newspapers, television, radio, and internet news outlets, actively seek stories of interest to their consumers. Excessive complaints, controversy, or unusual events such as storm emergencies will attract attention, and any resulting negative publicity can quickly galvanize public opposition to VM efforts. Without an explanation of the reasons for VM, media coverage is one-sided, public support flags, and the image of the utility can be negatively affected.

Most utility arborists are not accustomed to being the focus of media attention. Regardless of the circumstances, it is important for all personnel to understand that even innocent comments can be taken the wrong way and can adversely affect public perception. For this reason, all employees (except those authorized) must avoid making any substantive statements to media outlets, or posting on social media. Management must develop policies regarding who is authorized to make official statements, and direct personnel to refer inquiries through the correct channels.

Figure 7.5 These signs, located on distribution and transmission lines near Baltimore, Maryland, U.S., help educate the public about proper tree placement and the value of rights-of-way as wildlife habitat.

To ensure fair coverage, proactive efforts may be required, such as issuing press releases describing upcoming projects and their importance in improving the quality of service, and making authorized personnel available to explain these efforts. Again, demonstration projects and partnerships with environmental groups, as mentioned above, can be used as examples of good stewardship. In some cases, adjustments to the VM program specifications may be necessary.

Utility Regulatory Agencies

Most governments require that the cost and quality of critical services meet acceptable standards; for example, the California Public Utility Commission ensures "...the provision of safe, reliable utility service and infrastructure at reasonable rates, with a commitment to environmental enhancement and a healthy California economy" (California Public Utilities Commission n.d.). Government regulatory agencies usually consist of an appointed board with a support staff. These same agencies may also oversee other utility services such as natural gas, telecommunications, water, and transportation (e.g., taxi and motor carrier services).

Regulators and their staffs make decisions based on the information provided to them by concerned stakeholders, including companies being regulated and others who would be affected. For example, to fund a VM project, a utility may be required to justify a rate increase to regulators. Other stakeholders may adamantly oppose any increase in rates. The ability to provide and communicate accurate information about how utility VM programs benefit consumers is critical in such cases.

Suppliers

Suppliers include manufacturers of equipment and products used in providing VM services, and service providers, including firms and nonprofit organizations that provide training and **credentialing** services. To ensure that their needs are met, vegetation managers should maintain lines of communication and consider demonstration projects or other means of improving efficacy with suppliers of these products and services.

Land Management Agencies

In many areas, large tracts of land are managed by different regional or national government agencies, which often have widely disparate rules about how utilities may conduct VM activities. Aside from the logistical difficulty in changing methods and maintaining separate records for various line segments in different jurisdictions,

Public Relations, Customer Relations, Customer Service: What's the Difference?

Public relations, customer relations, customer service, and other similar terms are sometimes used interchangeably. While there may be some overlap in roles, these tasks often involve people from different departments, and even different companies, although all communicate with customers and ultimately play a part in customer satisfaction. It is also true that the actions of one group may have a major effect on the efforts of another.

Public Relations: An Overall Strategy

The Public Relations Society of America (PRSA) suggests that "**public relations** is a strategic communication process that builds mutually beneficial relationships between organizations and their publics" (PRSA 2012). In fact, many public relations personnel have relatively little direct contact with retail customers. Externally, utility public relations efforts have traditionally focused on enhancing overall customer perception of the company and its brand through media outreach, advertising, social responsibility, crisis management, and lobbying (Nye 1984). Public relations can also enhance communications within the company, for example, by ensuring that managers receive the information they need to make good decisions (Swann 2014).

Public relations personnel should become involved with VM projects when there is a higher level of customer concern or media attention, such as the reclamation of a transmission right-of-way, or a significant change to a distribution line-clearance program. For example, an effort to get a VM program back on cycle could require a high number of tree removals and more clearance than the public is accustomed to. In such cases, public relations professionals would devise a strategy to target affected stakeholders with appropriate information, and VM professionals would provide technical expertise.

Customer Relations: Information and Discussion

Customer relations involves providing information to customers, and discussion and negotiation as necessary. In VM, notification and work planning is often when the work is explained to customers, and where many questions and concerns are answered and addressed. The interval between notification and the actual performance of the work should be long enough for customers to have their questions answered, but not so long that customers forget that the work is scheduled to occur.

Customers vary in their response to VM. Some are very concerned, some are indifferent, others will react negatively under any circumstances, but many will base their response on how well the work is explained. Those involved with customer relations should recognize that more positive impressions will help shape a better overall public perception of the utility (Dike n.d.).

Customer Service: Getting it Done

In vegetation management, **customer service** includes getting the work done, and any follow-up that may be required. In most cases,

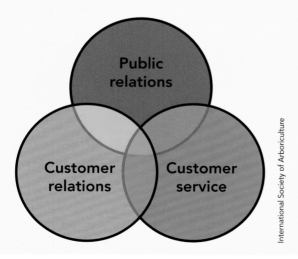

VM crews arrive to perform the work after customers have been notified and informed, although not all utilities follow this model. Sometimes work is scheduled without notice because of an emergency or a line construction project. Regardless, crew members should always be aware of basic considerations, such as appropriate crew appearance and behavior, and site clean-up. Crew personnel should be prepared to answer questions since customers may have additional concerns once the work has begun. How well crew personnel handle these situations is critical, because the wrong response can negate all previous efforts and result in a complaint or refusal.

the lack of continuity may negatively affect productivity and environmental performance. Efforts should be made to establish relationships with officials and to undertake demonstration projects and partnerships with willing partners that demonstrate the value and benefits of responsible ROW stewardship (UAA 2013).

Flexibility

In all cases, program managers should be willing to listen to various stakeholder concerns and, when circumstances warrant, consider accommodating stakeholder interests. VM activities are performed in myriad settings and may significantly impact environmental, cultural, and community resources. Aside from potentially violating government regulations that protect these resources, a lack of sensitivity can quickly generate opposition, cause negative publicity, and be far more expensive than accommodating the stakeholder concern in the first place. VM professionals should recognize when a certain amount of flexibility is necessary to avoid far more costly repercussions, while bearing in mind their primary obligation to ensure that the objectives of the VM program are accomplished.

OUTREACH METHODS

Information provided to customers in advance can have a significant effect on customer understanding and acceptance of VM activities (Kuhns and Reiter 2007). The **outreach** impact can be broadened by providing information in multiple formats, including different languages when necessary, to ensure that it is received and understood by as many customers as possible. Many common questions about what the work entails or why it is important can be answered in a basic informational brochure with easy-to-understand illustrations. Pictures and video can also be placed on a hand-held tablet in multilingual formats.

Printed Materials

Printed information such as brochures and pamphlets can be mailed directly, inserted into bills, left on customers' property, or handed out at community meetings and other events (Figure 7.6). However, creating and distributing printed material is costly, and there is no way to be certain that the information is reviewed or understood (Ladd 2010).

When used to inform customers about impending work, separate brochures should be developed for different kinds of work; for example, distribution utility pruning and transmission vegetation management should not be combined into the same notification brochure. This avoids confusion and makes it easier for the customer to understand what will happen on their property.

Printed information should be liberally illustrated and provided in a simple, easy-to-read format with relatively large font. Excessive or unnecessary technical detail, fine print, or multiple page handouts are less likely to be read and understood by customers (Ladd 2010). For customers who have questions or want more details, phone numbers, website addresses, and social media sites can be provided on the brochure.

Figure 7.6 This door hanger contains key information for customers about the work that is needed on their property, and who to contact with questions or concerns.

Email and Text Messaging

Most information that is provided in print format can be sent electronically by text message or email. These formats offer the advantage of being paperless, less costly, and more environmentally friendly. Links for more information can be placed directly in the text. Of course not all customers use these formats, or are willing to share their contact information, and databases must be maintained and updated as information changes. However, many customers prefer to receive information electronically and will opt to do so if given the opportunity (FirstEnergy n.d.).

Websites

An advantage of websites is that they allow more detailed information to be presented to customers. However, most utility websites provide an enormous variety of information, and there is little consistency in how utilities choose to organize and present web content, including VM information.

For customers with questions about VM, locating specific information on a utility website may be difficult. Also, there are many aspects of VM that may be presented. The information provided must address both utility and customer priorities, be easily accessed by mobile users, and be laid out so that customers can find information without wasting time or becoming frustrated.

To keep the customer's attention, information should be easy to find and should have the most commonly requested topics placed prominently. Frequently Asked Questions (FAQ) pages can be helpful in this regard (Ramirez n.d.). When the requested page opens, initial information should be well illustrated and presented in simple language. From there, links to more in-depth information can be provided for those who need it. This avoids overwhelming customers with too much information when just a simple explanation may be sufficient.

A "contact us" button can be added to websites to allow customers to send messages or questions directly to the utility, or if appropriate, a designated contractor representative. The person receiving public inquiries should have the capability to answer most questions and should know where to direct inquiries that may require further expertise. If a "contact us" link is provided, it is essential to reply to all requests (Ramirez n.d.).

Social Media

Social media provides an opportunity for ongoing, real-time engagement with customers on any number of topics, including storm response and VM. During emergencies, social media can be used to keep the public informed about restoration efforts or other critical information. At other times, utilities use social media platforms, including video uploads, to answer questions, have conversations, and provide relevant information, such as energy conservation tips, or where not to plant tall-growing trees (EL&P 2014).

It is important to ensure that information presented on social media is timely and accurate. This can be challenging for large, bureaucratic companies with many specialized departments that interact with customers (EL&P 2014). Each department has unique concerns, and many topics—such as VM—require technical expertise. Utilities should designate qualified personnel to update content on social media outlets, monitor posts, and respond appropriately to customer concerns.

Press Releases

Press releases provide specific information that the utility wants to make public. Often press releases are issued to get the utility's desired version of events out first and to preempt potentially negative publicity about an issue. If they are of interest, press releases are used by media such as newspapers, radio, TV, online services, and any other outlets where the public obtains information. Utilities routinely issue press releases for many reasons; however, VM is an area where proactive communication through press releases should be utilized frequently.

Press releases should be developed to provide information about specific VM projects, especially when public outcry is anticipated, such as "recovering " lines where pruning has not occurred for many years, when significantly changing the specification, or when beginning work in a neighborhood or community that has historically resisted utility VM efforts. They can also be issued to promote success stories, such as the completion of a large project, or to highlight the importance of VM. For example, if a study determined that 45 percent of outages were caused by tree failures, a press release could be issued to highlight this and explain how VM can improve electric service reliability.

Press releases should use information and imagery that is consistent with other outreach efforts whenever possible.

Community Meetings

Projects that will significantly impact local communities may require one or more community outreach meetings to inform the public about why the work is necessary and how it will be done (Figure 7.7). Information about the project and benefits to the community should be available. Working in coordination with utility public relations

Figure 7.7 Community meetings are an opportunity to provide concerned citizens with information about utility vegetation management projects.

professionals, an agenda, press releases (if necessary) and any information to be distributed should be developed ahead of time.

The people most likely to attend community meetings are a vocal minority opposed to the project being discussed. Proactively inviting participation from multiple affected stakeholders, with differing perspectives, ensures that a balanced view is represented (Enriquez 1997).

TALKING WITH CUSTOMERS

Many employees talk directly to customers, in person and on the phone. A professional approach will ease customer concerns and prevent small problems from morphing into costly complaints. Employees should expect to meet customers of different ethnic, linguistic, and socioeconomic backgrounds, and should always show respect, courtesy, and tolerance. The following are basic guidelines to ensure professional interaction with customers.

Make a Good First Impression

Customers often form opinions of employees and their companies within a few seconds. Customers who already have low expectations due to previous experience or word-of-mouth may be even more skeptical. Appearance and behavior that some employees might consider perfectly normal (e.g., spitting, smoking, teasing) may negatively influence the customer before any discussion begins; in fact, these are common and completely avoidable causes of complaints. While there are certain attributes that an employee has no control over, such as race, ethnicity or native language, there are many ways for employees to ease customer concerns from the start. Some tips for making a good first impression appear in Table 7.2.

Have a Good Attitude

Employees with a bad attitude have a negative effect on everyone around them, including fellow employees and customers. Moreover, they may not even realize that they are causing a problem. Managers should be aware of the attitudes of employees and how they may be affecting the performance of the entire operation (Barsade and Gibson 2007). The consistent delivery of good **customer service** requires employees to bring a fresh face to each customer and leave personal problems aside. This may necessitate a cultural shift across the entire operation.

Operate with Integrity

Companies and their employees are obligated to conduct themselves in compliance with the law and according to basic ethical principles in every aspect of their business. When the interaction involves customers or other stakeholders, everyone's reputation is on the line, including not just the employee, but the contractor, utility, and in fact, the entire industry. Creating a culture of honesty in an organization will encourage employees to follow suit (Ardichvili et al. 2009).

An employee should never take advantage of a customer for any reason, whether for personal gain or for that of their employer. Furthermore, employees have an obligation to report misconduct they have witnessed (ISA Certified Arborist Code of Ethics n.d.).

The impact of the work on a customer's property should never be deliberately downplayed or misrepresented, and any promises made must be

Table 7.2 Making a good first impression.

Positive	Negative
Supervisory/work planning personnel: collared shirts and clean pants, company logo or uniform	Clothing with tears, holes, and rips
Crew personnel: neat appearance, company shirt or uniform, appropriate personal protective equipment	Clothing with logos other than that of the contractor or utility (such as those of other companies or sports teams)
Identification if required	Clothing with offensive slogans or images
Vehicles clean, parked appropriately	Vehicles dirty or decrepit, blocking traffic, or parked haphazardly
Efficient traffic management, flaggers attentive and courteous	Poor traffic management, flaggers inattentive or indifferent
Respect for customers' property	Sleeping (or the appearance of sleeping) while taking breaks
Positive attitude and friendly approach	Spitting and urinating
Professional behavior, appropriate language	Smoking and flicking cigarette butts Foul language and name calling
Friendly greeting, including name of employee, company and utility, and reason for the visit	Indifferent or unfriendly tone, inadequate explanation offered for the visit

kept. This can easily happen unintentionally if, for example, notification personnel make a commitment that crew personnel are unaware of or are unable to keep. The result, in any case, is a dissatisfied customer.

But operating with integrity goes beyond honesty. Being on time for appointments, returning phone calls, and following-up on commitments are all basic expectations in customer service. Recording names, locations, contact information, dates, and times will ensure commitments are not overlooked.

Practice Active Listening

Active listening is a way to ensure that customers know that their concerns are being heard and understood by employees. People engaged in a discussion, especially when they are upset, are often focused more on formulating their own responses than actually hearing the other person's concerns (see the *Dealing with Angry Customers* section). When practicing active listening, an employee will deliberately:

- Set aside physical and mental distractions and focus attention on the customer.
- Make a genuine effort to hear what the customer is saying.
- Confirm that they are listening with affirming prompts ("uh-huh," "go on," "really!" etc.).
- Respond by restating the customer's concern in their own words (though without necessarily affirming the validity of the concern).

Using active listening demonstrates that the employee has heard the customer's concerns (Fitzpatrick and Fitzpatrick 2014). For example, after listening to a customer express concerns about a tree that was removed, an employee may say, "Yes, I see that without the tree the utility pole is clearly visible, and I understand how that must be very frustrating for you." This acknowledgement opens the door to further discuss the issue, including why it was necessary to remove the tree. Using active listening has many advantages, including:

- Ensuring that employees hear and are aware of customer concerns
- Encouraging discussion, which improves understanding on both sides
- Slowing the pace of the discussion, helping to avoid escalation or arguments, and allowing participants to make informed decisions

Anticipate Customer Concerns

Utility arborists talk to a broad cross section of people. While many are indifferent, others want specific issues addressed before they allow the work to proceed. Experienced utility arborists often anticipate customer concerns based on the neighborhood or type of property. For example, people with dogs, livestock, or other animals will want assurances that gates are properly closed and animals are protected from harm. Based on the individual situation, most concerns can be anticipated and resolved by actively listening and taking appropriate measures.

While most customers are reasonable, some have unreasonable expectations or make requests that cannot be accommodated. Using active listening, utility arborists should be able to discuss why this is the case, without being argumentative or offensive. There is also a small minority that will be difficult under any circumstances. During such encounters it is especially important to not argue or otherwise escalate the situation, and not to take things personally (see *Dealing with Angry Customers*).

Provide Good Answers to Questions

Questions from customers are an opportunity to explain and inform. Training employees to actively listen, empathize, and provide good answers will avert complaints and create satisfied customers (Figure 7.8). In fact, a commonly cited reason for customer dissatisfaction is an employee's lack of knowledge or unwillingness to answer basic questions about the service being provided (Amo n.d.).

Customers frequently ask why the work is necessary. A common mistake made by many utility arborists is to become defensive and blame the utility by answering, "We have to follow the utility's spec," or "We have to protect the utility's

Figure 7.8 Listening to customers and providing good answers to their questions reduces concerns and complaints.

power lines." Though these answers may be technically true, the benefits to the customer are ignored. What the customer hears is more like, "We have to protect *our* precious power lines from *your* lousy trees."

From the customer's perspective, a utility representative—whether a utility forester, work planner, supervisor, or crew member—should be able to answer questions about why the work is necessary and how it benefits the customer (Skill Standards for Utility Customer Service Representatives 2012). At a minimum, utility arborists should point out that the work is essential for the customers' safety and the reliability of their electric service.

Repeatedly answering the same or similar questions can lead to employee **desensitization**. It is important to remember that while the employee may have heard the same question many times, the customer knows nothing about the topic. Empathetic employees will keep in mind that most customers know little about this service or may view it as an unwelcome intrusion. Employees should do their best to provide customers with answers that are complete and easy to understand.

Most customers will be more cooperative when provided with good information. Some questions are more complicated and may require more nuanced answers, which employees can learn to provide with training and experience. Of course, some concerns require management involvement, but the vast majority of questions are easily answered by front-line personnel with some basic training (Pacific Northwest Center of Excellence for Clean Energy 2012).

For a list of common VM questions and suggested talking points, see *Customer-Friendly Language* on p. 211.

Make It About Them

Whenever possible, emphasize that *VM is an essential part of the utility's service that is provided for them at no extra charge!* Though a utility engineer may appreciate the need to protect utility property, a customer will be more interested in the *safety* of their property and the *reliability* of utility service for them, their neighbors, and the whole community. Better storm performance, including fewer and shorter power outages, and fire prevention are also good talking points, especially in areas prone to these events.

Phone Etiquette

Some customer communication will occur on the phone. Many of the techniques described above also apply during phone conversations, including the practice of active listening. In addition, employees should develop certain good phone habits, including:

- Answering all phone calls consistently, with a friendly tone, using their name and their company name
- Speaking clearly—be aware of the position of the microphone
- Avoiding distractions—if in a vehicle, pull off the road and turn off the radio
- Confirming that key information is heard and understood
- Returning phone calls at appointed times
- During conference calls, muting the phone to reduce interference from background noise

Customer-Friendly Language

When speaking to customers, choose positive and **friendly language** words with a clear meaning. Excessive use of jargon or an unfriendly tone can confuse and intimidate customers.

Jargon. It is easy to forget that language common in our industry can be confusing to others. The meaning of a "three-phase, 23 kV circuit" might be clear to a trained person, but it is gibberish to the average consumer. Likewise, to a customer, a "bucket" is a pail, a "lift" is a ride, and a "conductor" works on a train. Also, acronyms like "GF" (for general foreperson) are likely to be meaningless to a customer. Consider the context of the words used to describe work to be performed, and choose language that is easy for the customer to understand.

Friendly Words. Most customers don't like to hear words like "must," "can't," or "have to"—especially on their own property! When explaining work to customers, try to use friendlier words and phrases.

Some examples of unfriendly language and friendlier alternatives appear below.

Unfriendly tone	Suggested replacement
You have to let us remove that tree.	The tree is hazardous and really should be taken down.
No, we can't remove that tree.	We can clear the lines, but I am sorry, we're not able to remove that tree.
We can't get that brush until Tuesday morning.	We'll be back to get that brush first thing Tuesday morning.
You should just let us do our jobs.	I understand this is difficult, but it will be better for everyone if we move forward with this project.

Too much jargon	Suggested replacement
We're off-cycle on this circuit and the suckers have grown past the conductors.	There has been a lot of growth since the last time these trees were pruned.
The GF will order a bucket for that overhang above the three-phase.	My supervisor will get a lift truck to reach the branches above the power lines.

Dealing with Angry Customers

Inevitably, employees will encounter situations where customers have become angry, perhaps about the work, the actions of an employee, or for a completely unrelated reason. Regardless of the cause, or how upset a customer becomes, it is important for the employee to stay calm and in control. In these situations, an employee should avoid becoming personally involved or upset, and recognize that the customer's anger is usually not directed at the employee. Acting as if it were is likely to escalate the situation.

The use of active listening skills helps to acknowledge an angry person's feelings and calm the situation, even if the anger is misdirected or unreasonable—either way, the customer's anger is real. This can be done by:

- Demonstrating **empathy** by focusing attention on the customer and giving visual cues such as nodding agreement
- Making statements that confirm the person is being heard like "I get that" or "that's understandable"

The employee should allow the customer to "blow off steam." It is important not to be dismissive of the customer's concern, or to rush the situation. Saying things like, "I don't have to listen to that kind of language" or "This is not worth being upset about," is not recommended and may actually make things worse. At some point, an angry or upset person usually recognizes that they are acting badly and will calm down on their own—sometimes with an apology.

When it is certain that an error has been made, and the customer's grievance is legitimate, this should be acknowledged and handled with courtesy. Correcting a mistake can turn a bad situation into a success. Acknowledging mistakes, correcting them, and following up to make sure the customer is satisfied is great customer service.

It is also important to recognize when further discussion—or any discussion at all—is pointless, especially when there is any concern about safety or if the customer is threatening in any way, and when further discussion has no chance of improving the situation. In such situations, employees should not jeopardize their safety, and should quickly and quietly leave the property.

SUMMARY

Vegetation management, including transmission and distribution IVM, tree pruning and removal efforts, has often been underappreciated and undervalued by key decision makers and other stakeholders, from utility executives to regulators. Reasons are many, and include the public's lack of understanding of the value in terms of safety and reliability, a largely contracted workforce with high turnover rates, the relatively high short-term cost and long-term return on VM investment, and the cultural separation of VM, which is based in biological science, from the rest of the utility industry, which is oriented toward engineering and finance.

Utility vegetation managers should use a variety of communications media to ensure that information is received by various stakeholders, and focus attention on training front-line personnel in basic **customer relations** skills, including active listening. These efforts can improve both customer satisfaction and employee job satisfaction, resulting in better employee retention and reduced ancillary costs. Finally, targeted communication by all VM employees, from managers to front-line personnel, will improve relations and enhance understanding with internal and external stakeholders of both VM processes (such as IVM and utility arboriculture) and the value of VM in improving overall utility performance.

CHAPTER 7 WORKBOOK

Fill in the Blank

1. Most utilities list _____ _____ as a reason for maintaining vegetation, but customers often underestimate the risk.

2. People with an interest or concern in an organization and its decisions are known as _____.

3. Because of high turnover among contractors who perform vegetation management, safety and basic job skill training tends to have a higher priority than _____ training.

4. Issuing press releases that describe upcoming projects and their importance to quality service are _____ efforts that can offset negative media coverage.

5. _____ _____ is a way to ensure that customers know they are being heard and understood.

6. Repeatedly answering questions that are similar, such as those that may be asked of a contractor in the field, can lead to _____.

7. Customer-friendly language should keep _____ to a minimum because it can intimidate customers and confuse the message.

8. Some customers may not voice their displeasure about vegetation management directly with the utility company, but instead voice their opinion on social media, leading the utility to _____ customer dissatisfaction.

9. _____ _____ are often released to media outlets to provide specific information that the utility wants to make public.

Multiple Choice

1. Active listening involves all of the following *except*
   a. setting aside physical and mental distractions to focus attention on the customer
   b. making a genuine effort to hear what the customer is saying
   c. being completely quiet while the customer is talking to indicate that you are listening
   d. responding by restating the customers concerns in your own words

2. A utility CEO asks the line maintenance supervisor why so much money is being spent on line clearance when so few outages are caused by trees. Which of the following does this demonstrate?
   a. a well-run vegetation management program can be costly to the utility
   b. the value added by a vegetation management program may not be understood by all stakeholders within a utility
   c. the line maintenance staff was probably too large
   d. there were fewer than average tree-related outages in the area

3. You are a vegetation management contractor fielding complaints from a customer on a work site. Which of the following would be the best reply to a customer who questions why a tree is being pruned?
   a. "My work order says I have to do it."
   b. "This would be easier if you just let us get our work done."
   c. "This will help ensure uninterrupted service to the homeowners in this area."
   d. "I'm only a contractor; I didn't make this decision."

4. All of the following are ways for employees to make a good first impression with customers, *except*
   a. displaying proper identification
   b. wearing a clean company uniform
   c. parking vehicles haphazardly
   d. using appropriate language

5. What communication method is most effective for engaging with customers in real time about utility emergencies?
   a. printed materials
   b. phone calls
   c. social media
   d. community meetings

CHALLENGE QUESTIONS

1. Describe situations in which public relations personnel should become involved with vegetation management projects.

2. Detail the various outreach methods that can help utility vegetation managers communicate with customers and stakeholders, and when each method might be most effective.

3. Provide a more effective alternative for each of the following statements.

    a. I have no idea how long this will take.

    b. We have a right to park our truck here.

    c. That's what it's gonna look like if it hasn't been pruned in eight years.

    d. You won't be happy when your power goes out.

    e. Your neighbor called us.

GLOSSARY

A

AC: see *alternating current*.

accident: discredited term for unplanned, undesirable events that could result in unintentional injuries or property damage that are attributed to fate or bad luck (see *incident*).

accident pyramid: an illustration of Herbert Heinrich's principles of accident prevention, first presented in an example comprising untold thousands of unsafe acts or conditions at the base of the pyramid, which led to 300 close calls, 29 minor injuries, and ultimately, one major injury or fatality at the top.

accrual accounting: accounting methodology that records revenues and costs when they occur.

ACSR: see *aluminum conductor steel-reinforced cable*.

action threshold: a point at which the level of incompatible plant species, density, height, location, or condition threatens the stated management objectives and requires implementation of a control method(s).

active listening: a structured face-to-face communication technique that improves performance and increases understanding between people. The sender uses the name of the person with whom she or he is trying to communicate and states the message. The receiver paraphrases the message to indicate they understand the message and if the message was properly understood, the sender acknowledges it was.

acute toxicity: adverse effects resulting from exposure to a single dose or exposure to multiple doses of a substance over a short time frame (usually less than 24 hours).

aerial application: in vegetation management, application of herbicide using aircraft (helicopter or fixed wing aircraft).

aerial lift: a truck, logging skidder, or other vehicle mounted with a hydraulic boom-supported bucket or semi-enclosed working platform, used to elevate workers.

AHAS inhibitors: chemical class of commonly used integrated vegetation management herbicides in which the mode of action inhibits a key botanical enzyme, acetohydroxyacid synthase (AHAS), in the biosynthesis of critical amino acids.

airbrake switch: manually operated switch, usually deployed where two circuits meet. Often used to reroute electricity to a circuit or circuit portion from an energized feeder. They are not designed to automatically interrupt electric load or be used as line protection from overload, short circuit, or fault.

allelopathy: ability of some plants, including certain grasses, to release chemicals that suppress other plant species growing around them.

ALS inhibitors: chemical class of commonly used integrated vegetation management herbicides in which the mode of action inhibits a key enzyme, acetolactate synthase (ALS), needed for the biosynthesis of critical botanical amino acids.

alternating current (AC): electric current that reverses its direction many times per second at regular intervals, typically used in power supplies.

aluminum conductor steel-reinforced cable (ACSR): energized cable constructed from aluminum wire stranded around a steel core. Lightweight and relatively low-cost aluminum is often used as a conductor. However, it has a tendency to stretch when it heats, as it does when subject to high electric loads and elevated ambient temperatures. To minimize stretching and add strength, aluminum wire is stranded around a steel core in this cable construction.

aluminum core steel-reinforced cable: see *aluminum conductor steel-reinforced cable (ACSR)*.

ampere (amp, A): a measure of current. One amp is the charge carried by 6.25×10^{18} electrons moving past a given point in one second.

ANSI A300 (Part 7): *American National Standard for Tree Care Operations—Tree, Shrub, and Other Woody Plant Maintenance—Standard Practices (Integrated Vegetation Management a. Electric Utility Rights-of-way)* developed as one part of a ten-part series of industry-developed, national consensus standards of practice for tree care in the United States. Part 7 is specific to integrated vegetation management.

ANSI Z133: *American National Standard for Arboricultural Operations—Safety Requirements,* the industry-developed, national consensus safety standard of practice for tree care in the United States.

approach distances: when working near electric utilities, minimum distances that must be maintained between conductors and other energized equipment and qualified line-clearance personnel or persons other than line-clearance personnel and their bodies or tools. Approach distances vary with qualifications of personnel and with voltages.

automatic line recloser: a circuit breaker equipped with a mechanism that can automatically close the breaker after it has been opened due to a fault. They are designed to reset quickly to give the cause of a fault an opportunity to clear. They may immediately recharge the circuit several times at adjustable time delay—often two or three operations within seconds—before they open permanently.

auxin: plant growth regulator or substance that promotes or influences the growth and development of plants. Produced at sites where cells are dividing, primarily in the shoot tips. Auxin-like compounds may be produced synthetically.

auxin transport inhibitors: a class of selective herbicides that disrupt auxin transport. Auxin transport obstruction causes unnatural concentrations of auxin in meristematic shoots and roots, disturbing the auxin tolerances required for normal plant growth.

B

back feed: electric flow in an unintended direction through a process whereby de-energized lines become energized from an external source, such as a home generator.

bargaining unit: group of workers represented by a union local.

basal application: herbicide application made at the bottom 12 inches (30 cm) of the stem and root collar with an herbicide in an oil carrier.

behavior-based: a theory of industrial incident prevention pioneered by Herbert Heinrich that considers the overwhelming majority of incidents to be caused by unsafe acts.

biological methods: management of vegetation by establishment and control of compatible, stable plant communities using plant competition, allelopathy, animals, insects, or pathogens. Cover-type conversion is a type of biological control.

border zone: section of a transmission or pipeline right-of-way that extends from the wire or pipe zone to the right-of-way edge. The border zone is managed to promote a low-growing plant community of forbs, tall shrubs, and low-growing trees below a specified height (e.g., 25 feet [8 m]).

branch bark ridge: raised strip of bark at the top of a branch union, where the growth and expansion of the trunk or parent stem and adjoining branch push the bark into a ridge.

branch collar: area where a subdominant branch joins another branch or trunk that is created by the overlapping vascular tissues from both the branch and the trunk. Typically enlarged at the base of the branch.

branch removal cut: pruning cut that removes the smaller of two branches at a union or a parent stem, without cutting into the branch bark ridge or branch collar, or leaving a stub.

branch union: point where a branch originates from the trunk or another branch.

brand: name, service, trademark, color, or other attribute that distinguishes a product or service from competitors.

brush cutter: machine used to remove and grid saplings, small trees, and other incompatible vegetation.

budget: means to justify the allocation of resources, set priorities, and make decisions regarding what can and cannot be accomplished.

bus work: rigid conductor, usually rectangular or hollow tube design.

business continuity: the ability of an organization to maintain essential functions during, as well as after, a disaster has occurred.

C

capacitor: electrical device used to correct power factors. Temporarily stores a charge of electricity and returns it to restore line balance.

capacitive reactance: a charge that builds up between long parallel conductors. That portion of load slightly impedes current.

capital budget: monetary outlay for relatively large sums of money invested over multiple years.

cascading outage: a series of power interruptions on a transmission system whereby one interruption increases vulnerabilities and causes additional interruptions, which in turn cause further interruptions, potentially resulting in widespread blackouts.

chain of command: series of administrative ranks, positions, etc., each of which has direct authority over the one immediately below.

chemical control method: control of undesirable vegetation through the use of herbicide or tree growth regulators.

chemical pruning: method of pruning that utilizes certain herbicides to selectively control targeted branches, typically applied along the side of right-of-way corridors; also called *chemical side trimming*.

chronic toxicity: damage to an organ system by a substance through disruption of biochemical pathways or enzymes in areas other than the point of contact. The delayed damage resulting from repeated exposure to low amounts of a substance, evolving over a long period or a lifetime.

circuit: electrical network consisting of a conductor or series of conductors through which electricity flows, including a return path to the source.

circuit breaker: a switch that automatically interrupts the flow of electricity if a circuit becomes overloaded or another dangerous problem occurs.

circuit work: completing all necessary line clearance on a feeder or circuit basis.

clearance: (1) amount of open space between a tree or tree part and another object, the ground, or pedestrian or vehicle traffic on the ground; (2) in an electric utility system, or utility arboriculture, the distance between trees and utility lines (especially poles, crossarms, and conductors); (3) the creation or maintenance of these distances by qualified line-clearance personnel, line-clearance work.

closed chain of custody: handling and use of ready-to-use and diluted concentrate formulations in closed delivery systems.

codominant branches/stems: forked branches nearly the same size in diameter, arising from a common junction; may have included bark.

combined-cycle system: in electrical generation, a process that captures heat from the exhaust from generation and uses it to create steam to drive a second turbine.

common interest: in the context of this book, an interest between tree advocates and utilities in balancing the need for a reliable power supply with the cumulative benefits provided by urban forests. Examples include: maintaining existing trees to minimize risk to all urban infrastructure; maximizing environmental benefits, such as reduced urban heat islands and airborne particulates, and increased carbon sequestration; and creating attractive communities with a good quality of life.

compatible vegetation: vegetation that is desirable or consistent with the intended use of the site. For example, plant species that will never grow sufficiently close to violate minimum clearance distances with electric conductors.

comprehensive evaluation: at a program or project level, an accounting of all vegetation that could potentially affect management objectives, supplying a complete set of data upon which to base management decisions.

conductivity: capacity of a material to transmit electricity.

conductor: (1) in an electric utility system, metal wires, cables, and bus-bar used for carrying electric current—conductors may be solid or stranded (i.e., built up by an assembly of smaller solid conductors); (2) any object, material, or medium (e.g., guy wires, communication cables, tools, equipment, vehicles, humans, animals) capable of conducting electricity if energized, intentionally or unintentionally.

conductor sag: in an electric utility system, linear expansion of overhead wires in warm weather and/or during periods of high electrical flow, causing a droop or sag below its height at other times.

constraint triangle: depiction of the concept of triple constraint in project management, consisting of time, cost, and scope.

contract: a legally enforceable agreement between two or more parties. Every contract has five basic elements: offer, acceptance, consideration, legal and possible objective, and competent parties.

contractor: company retained by utilities to provide utility vegetation management services.

control methods: processes through which IVM managers achieve the objective of maintaining a desirable plant community. Includes manual, mechanical, chemical (herbicide and tree growth regulators), biological, and cultural control methods (control options)—used alone or in combination.

coppicing: silvicultural technique in which trees are cut at the ground to encourage growth of sprouts to be used for various purposes. Sometimes described as basal pollarding.

Coriolis force: effect that causes deflection of moving objects to the right in the Northern Hemisphere and to the left in the Southern Hemisphere due to the Earth's rotation.

corrosive: a substance that damages animals (including humans) at the point of contact.

cost center: accounting configuration for expenditures that do not generate income.

cover-type conversion: type of biological control in which a plant community is converted from incompatible to compatible species using selective techniques that provide a competitive advantage to short-growing, early successional plants, allowing them to thrive and successfully compete against unwanted tree species for sunlight, essential nutrients, and water.

cover-type mapping: technique that uses aerial photographs followed by ground checks to determine the nature of plant communities on a site.

credential: ISA certification or other industry qualification, certifying that an individual has attained a generally accepted level of knowledge in a specific subject area or profession, such as tree work.

crew: in utility arboriculture, the basic unit of personnel and equipment, consisting of two to five or more people, usually led by a foreperson. Crews are often characterized by their specialized skills and types of equipment used, e.g., manual, lift, off-road, etc.

crisis management: reacting to conditions rather than working toward long-term objectives. Crisis management is characteristic of a program with poorly defined management strategies, ineffective management, underfunded or understaffed program, or an undereducated manager or workforce (Miller et al. 2015).

critical path: in project management, the longest period of time a series of tasks requires from project beginning to end.

critical service provider: an organization, such as a utility, communications company, bank, etc., whose incapacitation or destruction would have a debilitating effect on public interests.

cultural control method: control method that modifies habitat to discourage incompatible vegetation and establish and manage compatible plant communities.

current: the flow of electricity.

current surge: inrush of electricity caused when electrical devices, such as motors or transformers, are first switched on, at which time they can draw several times their normal operating electrical load; often occurs when circuits are first re-energized following an outage.

customer relations: discussions and negotiations with individual customers.

customer service: attention to customers' interest before, during, and after vegetation management work.

cutout: a combination fuse and switch used in overhead primary lines to protect distribution transformers from current surges and overloads. An overcurrent caused by a fault will cause the fuse to melt, disconnecting the transformer from the line. It can also be opened manually by utility linemen.

cycle: in vegetation management, the planned interval between scheduled maintenance established to maintain adequate clearance between trees and utilities.

cyclone: a large area of low atmospheric pressure, characterized by inward-spiraling winds.

D

decision package: a proposal or plan that outlines project expenditures, outputs, and alternatives, which receive funding in order of priority.

defects: in tree risk assessment, injuries, growth patterns, decay, or other conditions that reduce the tree's structural strength.

deferring maintenance: delaying work, usually for financial reasons.

delta construction: electric distribution configuration with three phases, but no common neutral wire. There is no fixed difference between phase-to-phase and pha2se-to-ground voltages.

dependencies: relationships that dictate when tasks in the work breakdown structure begin and end.

derecho: complex of thunderstorms that travels more than 240 miles (386 km), with wind speeds greater than 58 mph (93 km/h).

desensitization: process by which personnel become indifferent or disconnected from customers and their concerns.

direct contact: electrical contact that occurs when someone touches an energized fixture.

directional pruning: selective removal of branches to guide and/or discourage growth in a particular direction.

disaster: natural or man-made occurrence disrupting the normal conditions of existence and causing a level of suffering that exceeds the capacity of adjustment of the affected community.

disaster declaration: proclamation by a government that releases resources (e.g., funds, personnel, or special procedures) or otherwise hastens the response to a disaster (see *state of emergency*).

disconnect switch: manual device that isolates electric lines or circuits, allowing visual confirmation that it is open. It is not a protection device, but is used to isolate areas undergoing an outage or reroute electricity from an adjacent energized circuit or circuit section to an affected area to limit the extent of an outage.

distribution lines: in an electric utility system, electric supply lines usually energized between 2.4 and 23 kV (contrast with *transmission lines*).

distribution network: in an electric utility system, the portion that delivers electricity to end-users. Includes distribution substations, primaries, distribution transformers, secondaries, and service lines.

dose: the quantity of a substance absorbed into the body.

down guy: a guy wire secured at ground level.

E

early successional plant community: plant community dominated by annual and perennial herbaceous plants that develops soon after disturbance. In a utility context, cover-type conversion inhibits successional progress past an early stage.

easement: legal, non-possessory interest in real property that conveys use or partial use, but not ownership, of all, or more typically a portion, of an owner's property.

electric grid: interconnected network for delivering electricity from suppliers to consumers, including generation, transmission, and distribution.

electric shock: physiological reaction to the passage of electrical current through the human body.

electric utility: (1) corporation, person, agency, authority, or other legal entity aligned with distribution facilities for delivering electricity to consumers; (2) the system of facilities maintained by an electric utility to carry out its operation.

electrical conductor: (1) in an electric utility system, metal wires, cables, and other system components used for carrying electric current—conductors may be solid or stranded (i.e., built up by an assembly of small solid conductors); (2) any object, material, or medium (e.g., guy wires, communication cables, tools, equipment, vehicles, humans, animals) capable of conducting electricity if energized, intentionally or unintentionally.

electrical potential (voltage): difference in electric energy between an energized conductor and another energized conductor, the ground, or other object.

electrocution: death from electrical shock.

electromagnetic field: force field comprising electric and magnetic elements associated with electric charge in motion.

electromotive force: voltage.

empathy: the action of understanding, being aware of, being sensitive to, and vicariously experiencing the feelings, thoughts, and experience of others.

energized conductor: electrical conductor through which electrical current is flowing. This may be intended and expected or the unintended and unexpected result of an electrical short.

Enhanced Fujita Scale: six-level numerical, damage-based classification of estimated wind speeds, usually applied to tornadoes.

epicormic shoot: shoot arising from a dormant bud or from newly formed adventitious tissue.

EPSP inhibitors: non-selective herbicides (e.g., glyphosate) that inhibit 5-enolpyruvylshikimate-3-phosphate (EPSP) synthase, an enzyme without which a plant cannot produce some botanical proteins required for growth.

evergreen contract: contract without a set expiration date.

excurrent: pattern of tree branching characterized by a central leader and a pyramidal, cone-shaped crown (contrast with *decurrent*).

exposure: amount of chemical reaching the body.

external communication: dialogue with residential and commercial customers, as well as media outlets, governments, landowners, and regulatory agencies. Communications outreach efforts should be developed and implemented for each type of external stakeholder.

extratropical cyclone: cyclonic-scale storm that is not a tropical cyclone, usually referring to cyclones of middle and high latitudes.

F

FAC-003: Transmission Vegetation Management reliability standard for vegetation management along transmission lines, released by the North American Electric Reliability Corporation (NERC).

failure mode: in the context of this book, vegetation-caused electrical outages. There are two types of vegetation-related failure modes: mechanical tear down and electrical short circuit.

fate in soil: the ultimate behavior of an herbicide in the soil, including its absorption, potential for erosion, half-life, and degradation.

fault: unintentional and undesirable conducting path or blockage of current in an electrical system.

feeder: (1) in an electric utility system, a high-priority electric distribution supply line—generally 12,000 to 34,500 volts—that carries electricity from distribution substations to other primary distribution supply lines or distribution transformers; (2) distribution circuit.

field winding: (1) coil of wire through which a current passes in an electric generator or motor; (2) wire winding in a stator.

first responders: personnel likely to be the first to arrive and assist in emergencies (including storms and other natural disasters) such as police, firefighters, paramedics, and others with specialized skills and equipment as needed.

freezing rain: rain that falls in liquid form but freezes upon impact to form a coating of glaze upon the ground and on exposed objects.

friendly language: choice of words that deliberately emphasizes a helpful or non-confrontational approach. For instance, instead of saying "you have to…," using "it would be better if we…."

frill treatment: method of herbicide treatment involving herbicide application into cuts in the trunk. Also called *hack and squirt*.

frond: large, divided leaf structure found in palms and ferns.

funnel cloud: rotating cloud associated with a cumuliform cloud, not in contact with the ground. Can be regarded as a forming tornado.

fuse: protective device designed to melt above a specified voltage, thereby disconnecting a circuit.

G

gas turbine: in electrical generation, steam turbine fueled by natural gas.

geographic information system (GIS): layered mapping application consisting of a base map and overlays, containing features and information that can be viewed singularly or in combination.

GIS: see *geographic information system*.

global positioning system (GPS): satellite-based application used for navigation and to gather field data, including locations of structures, vehicles, people, and geographic features.

GPS: see *global positioning system*.

fleet tracking: tracking the location and movement of vehicles using a combination of signals from GPS satellites and cellular towers (see *vehicle monitoring system*).

Graham cover-type mapping: establishing a cover-type map using aerial (satellite) imagery in combination with tree species identification along transect lines or at randomly or systematically selected points.

grid work: distribution cycle work performed within a specified geographic boundary.

ground wire: common return path for an electrical circuit or physical connection to the earth.

grounded: electrically connected to the earth, providing a path for the flow of electricity to prevent accidental energizing.

grounding: (1) *n.* the act of creating a path for an electrical current to reach the earth (**earthing**, in British English); (2) *adj.* having a path for an electrical current to reach the earth (**earthing**, in British English).

ground-to-conductor clearance: in an electric utility system, the clearance or distance between live or energized conductors and the ground. In the United States, these clearances are specified in the National Electrical Safety Code (NESC).

guy wire: wires used for support in places such as where a line turns or dead-ends, in areas of high winds or at other anticipated stress points.

H

hack and squirt: see *frill treatment*.

half-life: amount of time it takes half of the quantity of a substance to dissipate, indicating herbicide persistence in this text.

hazard tree: tree that has been assessed and found to be likely to fail and cause an unacceptable degree of injury, damage, or disruption. Hazard trees pose a high or extreme level of risk.

heading cut: cutting a shoot or branch back to a bud, stub, or small lateral branch. Cutting an older branch or stem back to a stub in order to meet a structural objective.

heavy, wet snow: snow with a water equivalent ratio of 10:1 or less.

herbicide: specialized chemical substance used to kill plants by interfering with critical botanical biochemical pathways.

Hertz (Hz): in electric generation, the number of cycles, or revolutions of spinning generation magnets, per second. Alternating current periodically varies in amplitude in conjunction with spinning generation magnets, intensifying from zero to a maximum value in a positive direction back to zero, and minimum value in the negative direction before returning to zero. The changing intensity and direction of alternating current is graphed in a sine wave. The time it takes for a magnet to complete a revolution is a cycle.

high-reliability organization: group that takes a systems approach to safety, recognizing that humans are imperfect and that errors have to be considered in the context of the system in which they occurred.

high-voltage lines: in an electric utility system, lines with voltages greater than 750 V.

hotspotting: process by which limited tree work is scheduled on a specific area of the network where shorts or outages are occurring.

hurricane: regional term of the Western Hemisphere for a tropical cyclone with wind speeds 74 mph (119 km/h) and greater.

hydration: condition of having adequate fluid in the body tissues.

hydroelectric: electric generation using water.

I

ICS: see *Incident Command System*.

imminent threat: a vegetation condition that could cause damage or interruption of service to overhead energized facilities or pipelines at any moment.

impedance: the sum of resistance—inductive resistance and capacitive resistance.

incident: unplanned, undesirable event that could result in unintentional injuries or property damage.

Incident Command System (ICS): standardized on-scene incident management approach that allows responders to adopt an integrated organizational structure equal to the complexity of and that meets the demands of a single incident or multiple incidents, without being hindered by jurisdictional boundaries.

incompatible plant species: plant species that are inconsistent with the use of a site (e.g., tree species sufficiently tall that they will grow into overhead conductors during their life span).

indirect contact: touching a conductive object in contact with an energized fixture.

inductance: electrification of a wire by passing it through a moving electromagnetic field.

inductive reactance: reactance is caused by a small amount of load lost when a current flows into a coil (such as those in transformers).

initial clearing: from a vegetation management perspective, clearing a right-of-way prior to construction, or the first work clearing an overgrown right-of-way.

insulator: material with poor conductivity—such as fiberglass, glass, polymers, and porcelain—used to separate energized conductors from poles or other objects or structures that must not be energized.

integrated pest management: the use of all available pest control methods to keep pest populations below acceptable levels. Methods include cultural, biological, chemical, physical, and genetic.

integrated vegetation management (IVM): system of managing plant communities based in integrated pest management (IPM), in which managers identify compatible and incompatible vegetation, consider action thresholds, evaluate control methods, and select and implement controls to achieve specific objectives. The choice of control methods is based on the anticipated effectiveness, environmental impact, site characteristics, safety, security, economics, and other factors.

interconnection reliability operation limit (IROL): the value (such as MW, Mvar, amperes, frequency, or volts) derived from, or a subset of the System Operating Limits, which if exceeded, could expose a widespread area of the Bulk Electric System to instability, uncontrolled separations or cascading outages.

intermittent fault: repeated momentary interruptions in the same place due to the same cause.

internal communication: dialogue between departments and personnel that is concerned with service reliability, power quality, maintenance costs, and public relations.

interruption: disruption of electrical supply.

IROL: see *interconnection reliability operation limit.*

irreversibility: effect of chemical exposure that cannot be changed or remedied.

IVM: see *integrated vegetation management.*

J

job briefing: meeting conducted at the job site by the employee in charge of the work that focuses on the site-specific hazards associated with the work to be performed.

just-in-time management: results-based strategy designed to target trees as close as possible before they interfere with facilities. It is a modification of hotspotting insofar as it systematically covers an entire system in a specified time, but only addresses trees that are anticipated to contact electric lines over that cycle.

K

key performance indicator (KPI): contractual target against which a contractor is evaluated. It can include unit price components (e.g., cost per tree, acre, or mile), as well as elements, such as work quality, customer satisfaction, reliability, safety, certifications, or other factors.

kilovolt (kV): 1,000 volts. Unit used for the measurement of electrical force.

KPI: see *key performance indicator.*

L

label: document accompanying an herbicide package in a standard format including an ingredient statement, Environmental Protection Agency (EPA) registration and establishment numbers (in the U.S.), compatibility with other chemicals, recommendations to prevent overdose, hazards to wildlife, first aid statements, storage and disposal, as well as protective clothing and equipment advisories. Describes the approved uses for the pesticide and directions for its safe application. In the U.S., the law forbids use inconsistent with the label.

labor union: organization of employees that collectively negotiates wages, benefits, working conditions, and other matters.

lagging indicators: serious injuries or fatalities.

large-scale response: storm response requiring movement of crews from outside the immediate vicinity, often for an extended period, and therefore including the need for temporarily housing and feeding response personnel.

lateral: secondary or subordinate branch or root.

lateral bud: vegetative growth point on the side of a stem (contrast with *terminal bud*).

LC_{50}: inhalation concentration of a substance that kills half the subject test animals within two weeks.

LD_{50}: oral or dermal lethal dose of a substance that kills half the test animals within the two-week test period.

leading indicators: unsafe acts and close calls at the bottom of the accident pyramid.

level one assessment (limited visual assessment): a visual assessment from a specified perspective of an individual tree or a population of trees near specified targets, conducted in order to identify obvious defects or specified conditions (Dunster et al. 2017).

level two assessment (basic assessment): detailed visual inspection of a tree and surrounding site that may include the use of simple tools.

lever arm: length of a branch or trunk to be considered in evaluating the combination of forces that could lead to failure.

LiDAR: acronym for light detection and ranging, technology that uses laser pulses to evaluate field conditions and workloads on rights-of-way, with results depicted in a computer image.

lightning arrestor: device installed to protect electric facilities from voltage surges caused by sudden electrostatic discharges during electrical storms (lightning).

line: distribution or transmission electric facility including wire, poles, and attachments.

line sectionalizer: protective device mounted on distribution poles or crossarms that isolates line sections or protective zones in order to limit the number of customers who lose service as the result of a fault. They cut off current when an upstream recloser operates, isolating a fault in the line beyond the sectionalizer.

lion tailing: poor pruning practice in which an excessive number of interior lateral branches are removed, resulting in a concentration of growth at branch ends.

load: force applied to a structure, such as a tree.

lockout: permanent operation that requires manual closure of automatic line reclosers.

low-voltage lines: (1) overhead or underground electric supply lines up to 750 volts, typically 120 to 480 volts, used to deliver electricity to end users (see *secondary*, *distribution line*, and *service drop*); (2) other electric supply lines with operating voltages typically at or below 24 volts, such as landscape lighting, alarm systems, signaling systems, and telecommunication systems (contrast with *high-voltage lines*).

lump-sum contract: contractual strategy where vendors submit a single price for a set amount of work (e.g., project or the entire account for a specified period of time).

M

maintenance/management interval: planned length of time between vegetation management activities. Also called *maintenance cycle*.

manual control method: vegetation cutting or removal using tools carried by hand.

mechanical control method, mechanical pruning: vegetation removal using heavy equipment fitted with power saws or other cutting devices, such as saws mounted on booms or suspended from a helicopter.

media relations: linkages with the media (print, TV, radio, online, personalities, and resources) that facilitate an organization in getting favorable, timely, and widespread editorial coverage.

megawatt: one million watts.

meristem: undifferentiated plant tissue in which active cell division takes place. Found in the root tips, buds, cambium, cork cambium, and latent buds.

messaging: strategic communication targeted at stakeholders.

mid-cycle pruning: process where trees growing significantly faster than most other trees are pruned at approximately half the scheduled maintenance cycle length, enabling a longer interval for slower-growing trees.

minimum approach distance: Established by ANSI Z133, the distance from energized conductors inside of which qualified line-clearance arborists, incidental line-clearance arborists, or persons other than qualified line-clearance arborists may not work. Distance increases with increasing voltage and varies with training of personnel.

minimum vegetation clearance distance (MVCD): calculated minimum distance between conductors and vegetation to prevent spark-over, for various altitudes and operating voltages, that is used in the design of transmission facilities.

mobilization: process of assembling and preparing employees, equipment, and supplies for deployment during a storm emergency.

mode of action: botanical biochemical pathways disrupted by herbicides; common types include ALS or AHAS inhibitors, synthetic auxins, EPSP inhibitors, photosystem I inhibitors, photosystem II inhibitors, proton inhibitors, and auxin transport inhibitors.

momentary interruption: transient fault in an electrical system, lasting from 33 to 133 milliseconds.

monocot: see *monocotyledon*.

monocotyledon: plant with an embryo that has one single seed leaf (cotyledon). Examples are grasses and palms.

multiple causation theory: refinement of the behavioral-based safety theory that considers workplace injuries to be caused by *a* number (rather than *the* number) of contributing factors and causes, which randomly interact.

mutation: genetic change.

mutual assistance program: in utility arboriculture, pre-arranged cooperation between utilities to ensure a certain level of response is available in the event of emergencies.

MVCD: see *minimum vegetation clearance distance*.

N

National Electrical Safety Code® (NESC): standard in the United States covering basic provisions for safeguarding persons from hazards resulting from installation, operation, or maintenance of conductors and equipment in electric supply stations, overhead and underground electric supply, and communication lines. It also contains work rules for construction, maintenance and operations of electric supply, and communication lines and equipment.

National Labor Relations Act: primary law governing collective bargaining in the United States.

natural pruning: process of branch removal in which the pruning cuts are made at nodes and in relation to the positions of the branch collar and branch bark ridge.

NERC: see *North American Electric Reliability Corporation*.

NESC: see *National Electrical Safety Code®*.

neutral wire: common return path for an electrical system.

news media: all sources of news and information, including: TV, radio, newspapers, magazines, web pages, and blogs.

node: slightly enlarged growth point on a stem where leaves and buds arise (contrast with *internode*).

non-damaging level: established, acceptable point (economic, programmatic, or aesthetic) below which pests interfere with objectives. In the context of IVM, the point at which incompatible plants are allowed in a right-of-way.

non-selective herbicides: herbicides that affect a broad range of plant species.

North American Electric Reliability Corporation (NERC): not-for-profit international regulatory authority whose mission is to assure the reliability of the bulk power system in North America.

Occupational Safety and Health Act (OSHA): in the United States, the legislative act dealing with health and safety in the workplace. In Canada, the federal, provincial, and territorial jurisdictions each have their own legislation. In Australia, there is a Work Health and Safety Act in each state.

Occupational Safety and Health Administration (OSHA): in the United States, the federal agency responsible for establishing and enforcing safety work rules. Individual states may also have OSHA agencies. In Canada, the responsible agency is the Occupational Health and Safety Administration (OHSA).

ohm (Ω): measure of electrical resistance consisting of one volt flowing at one amp.

oil switch: a manually operated switch used on distribution lines to disconnect heavy electric loads. It is capable of extinguishing an electric arc created by interrupting a heavy load.

oncogenicity: the capacity to cause cancer.

operating budget: budgets that finance day-to-day activities, including salaries, rent, utilities, and supplies.

operation: in an electric transmission or distribution system, the activation of a protective device.

OSHA: see *Occupational Safety and Health Act, Occupational Safety and Health Administration*.

outage: interruption to the electrical supply.

outreach: the activity or process of bringing information or services to people.

overcurrent: amperages in a conductor that are larger than those for which it is rated.

performance appraisal: employee evaluation.

performance budgeting: a budgeting strategy that stresses the goals and objectives of a program. It often involves mission statements and performance measures.

performance-based contract: results-oriented method of providing a service that focuses on the outputs, quality, or outcomes, including the achievement of specific, measurable performance standards and requirements. These contracts may include both monetary and non-monetary incentives and disincentives.

peripheral zone: in vegetation management, the area outside the right-of-way, where tall-growing species may be allowed, although they need to be monitored for risk.

phase (Ø): in an electric utility system, a single primary energized conductor affixed to a pole or cross arm. Technically, phase refers to the synchronized movement of electrical energy and pertains to common designs of electric supply lines, such as single phase (one primary conductor), two phase (two primary conductors), or three phase (three primary conductors).

phase-to-ground: in an electric utility system, the electrical potential between an energized conductor and the ground.

phase-to-neutral: in an electric utility system, the electrical potential between an energized conductor and the neutral wire.

phase-to-phase: in an electric utility system, the electrical potential between two energized conductors or between an energized conductor and a neutral conductor.

photosystem I inhibitors: group of IVM herbicides (e.g., paraquat and diquat) that intercept electrons from photosystem I in photosynthesis. Those electrons initiate a self-perpetuating biochemical lipid oxidation chain reaction that results in cell membrane degradation, which leads to leaf wilting and desiccation.

photosystem II inhibitors: group of IVM herbicides that that block electron transport in photosystem II in photosynthesis, preventing it from producing the carbohydrates needed for plant growth. Biochemical reactions occur that attack lipids and proteins, deteriorating cell walls, which results in leaky membranes, causing cells to dry and disintegrate.

pipe zone: area of a utility pipeline right-of-way over the pipe and extending out both sides to a specified distance.

point sampling: method of sampling a geographical area by selecting points in it, especially by choosing points at random on a map or aerial photograph.

poison: a substance with high acute toxicity. In the U.S., substances that have an oral LD_{50} fewer than 50 mg per kg or a dermal LD_{50} fewer than 200 mg per kg of body weight are categorized as poisons. The labels for poisonous materials are marked with a skull and crossbones. Herbicides commonly used in IVM are not poisons.

pollarding: specialty pruning technique in which a tree with a large-maturing form is kept relatively short. Starting on a young tree, internodal cuts are made at a chosen height, resulting in sprouts. Requires regular (usually annual) removal of the sprouts arising from the same cuts. Callus knobs develop at the cut height from repeated pruning.

predictive maintenance: Type of preventive maintenance performed continuously or at intervals, using field data to diagnose and monitor a condition or system.

pre-inspector: employee assigned to gather information, notify customers, and identify specific required tasks ahead of scheduled maintenance operations.

press release: article issued to the media describing upcoming projects and their importance in improving the quality of service.

pre-staging: movement of personnel and equipment into place prior to a storm strike.

preventive maintenance: maintenance operations scheduled to take place in advance of failure or other problems (contrast with *reactive maintenance*).

primary line: in an electric utility system, high-voltage distribution electric supply line, usually 2,400 to 34,500 volts, which carries electricity from distribution substations to distribution transformers.

primary voltage: in an electric distribution system, electric potential above a designated level, generally upstream of transformers.

profit center: accounting configuration where expenditures are sources of revenue. A profit center can be a product or service, a region, a distribution channel, or other consideration.

program budgeting: budgeting strategy in which line items are organized into programs, rather than accounts, with a focus on outcomes, rather than inputs. Examples of programs might be labor, chemicals, or individual projects.

pruning cycle: in utility arboriculture, the time scheduled between pruning events that is established as a guideline for accomplishing stated program objectives.

pruning machine: a vehicle mounted with a telescoping boom and fitted with a circular saw head.

pruning system: process used to achieve the desired long-term form of a plant.

public relations: efforts that help an organization and its publics adapt mutually to each other.

Q

quadrat sampling: statistical sampling method that uses replicated units (quadrats) of a designated size to determine the abundance of plants or other organisms in a designated area.

qualification: a designation earned by an individual following demonstration, through testing or other performance measures, of sufficient expertise in a specific field of study.

qualified line-clearance arborist: an individual who, through related training and on-the-job experience, is familiar with the equipment and hazards in line clearance and has demonstrated the ability to perform the special techniques involved. This individual may or may not currently be employed by a line-clearance contractor.

R

radial system: in electric distribution systems, a simple design with only one path of power flow, consisting of a single track of three-phase lines from a substation, with three or single-phase laterals tapping off them to serve end users.

reactive maintenance: maintenance operations that take place after a problem has occurred (contrast with *preventive maintenance*).

reduction cut: pruning cut that removes the larger of two or more branches or stems, or one or more codominant stem(s), to a live lateral branch, typically at least one-third the diameter of the stem or branch being removed.

resistance: the ratio of voltage to current, measured in ohms. Resistance is analogous to friction or drag.

response growth: new wood produced in response to loads to compensate for higher strain in marginal fibers; includes reaction wood (tension and compression) and woundwood. Also called *compensatory growth* or *adaptive growth*.

restoration: process of repairing and re-establishing critical utility services to end users during utility storm response.

restoration pruning: pruning to redevelop structure, form, and appearance of topped or damaged woody plants.

results-based management: approach to work planning that seeks to minimize tree-caused interruptions by focusing on areas of greatest need, rather than set time periods.

returnable, reusable (R/R) container: chemical container designed for multiple uses, typically made of recyclable, translucent, chemically resistant plastic with a service life of roughly 5 years or 30 return cycles.

reversibility: effects of chemical exposure that are not permanent and can be changed or remedied.

right-of-way (ROW): defined area of land, usually a linear strip, reserved for the passage of traffic (e.g., paths and roadways) or the construction, maintenance, and operation of various aboveground or underground utilities. Rights-of-way may be granted by easement rights and may cross a single property or many properties (highways, railroads, or utility corridors are common examples).

roster: list of the people or things that belong to a particular group, team, etc.

roundover: discredited pruning technique whereby trees are severely reduced to a predetermined shape using heading cuts.

ROW: see *right-of-way*.

S

safety committee: group of employees responsible for developing safe work practices.

safety culture: a shared group of values or accepted social norms among workers regarding safety.

safety stand-down: a pause in operations, often in response to a specific incident or theme, for employers to talk directly to employees about safety.

Saffir-Simpson Scale: classification scheme for hurricane intensity based on the maximum surface wind speed and the type and extent of damage done by the storm.

SCADA: see *supervisory control and data acquisition*.

SAIDI (System Average Interruption Duration Index): a measure of the *number of minutes* the average customer is out of power over a year's time (or any specific time period): [total duration of customer interruptions/total number of customers].

SAIFI (System Average Interruption Frequency Index): a measure of the *number of outages* experienced by the average customer over a year's time (or any specific time period): [total number of customer interruptions/total number of customers].

secondary (secondary lines): in an electric utility system, a lower-voltage (generally 110 to 750 volts) electric supply line that carries electricity from distribution transformers to service lines.

secondary voltage: voltage leaving a transformer in the intended direction. On distribution systems in North America, secondary voltage is 120 to 240 volts.

seed bank: native, compatible species lying dormant on-site.

selective foliar application: spraying leaves and shoots of specific target plants; can be either low or high volume.

selective herbicides: herbicides that are limited to the control of specific kinds of plants when applied according to the label (e.g., synthetic auxins, which are a class of selective herbicides that kill broadleaved plants, but do not harm grasses).

service container: Smaller capacity receptacles that can be refilled by applicators from supply containers. They typically have 2.5 to 5 gallon (10 to 20 liter) capacities, and should be sufficiently durable for repeated use. They are designed to hold the same amount of custom blend formulations as the application containers to minimize the need for spray crews to perform careful measuring in the field.

service drop: in an electric utility system, a distribution secondary line that carries electricity from a transformer to a home or business.

service interruption: disruption of utility service due to storms, equipment failure, human error, or other causes.

service line: see *service drop*.

set objectives: the initial phase of the integrated vegetation management model.

shearing: (1) cutting leaves, shoots, and branches to a desired plane, shape, or form, using tools designed for that purpose, as with topiary; (2) whole tree removal with devices mounted on excavators or other heavy equipment.

shear: in mechanics, the movement or failure of materials, especially laminar material such as wood, by sliding side-by-side.

short circuit: current that is bypassing a designed conducting path.

side-pruning: removal of branches on one side of the tree to provide clearance from utility infrastructure, traffic, or buildings.

simple cash accounting: recording financial transactions when they are paid.

sine wave: graph of the changing intensity and direction of alternating current.

site conditions: factors unique to a particular location (e.g., soils, exposure, slope, etc.) that should be taken into account when assessing the likelihood of a tree or branch failure.

sleet: rain that freezes into solid ice pellets before hitting the ground.

small-scale response: storm response handled entirely by locally based personnel.

SMART goals: in project management, a means of setting and evaluating goals using the elements of the acronym: specific, measurable, achievable, relevant, timely.

social media: web-based sites that enable individuals, organizations, or corporations to share information, ideas, photos, and videos with networks or the public.

soil drench: direct ground application of chemical growth regulators or pesticides, usually at the base of a targeted tree.

soil injection: placement of chemical growth regulators or pesticides into the ground using a specialized tool that provides a metered dose.

solar generation: electrical generation using the sun's energy.

span guy: a guy supporting one pole anchored to a second pole.

spark-over: luminous discharge of electricity through a gap between two conductive objects (e.g., a power line and a tree).

specifications: document stating a detailed, measurable plan or proposal for provision of a product or service.

staging area: designated location where personnel and equipment are gathered and prepared for deployment.

stakeholder: person or group that has an interest in, or is affected by, an activity or decision. External stakeholders are outside a business or project. Internal stakeholders work within an organization or project.

state of emergency: extraordinary circumstance, usually declared by a government, in which normal rules and procedures are suspended and unusual measures may be taken, often to avert or mitigate a disaster.

stator: stationary part of an electrical generator. Contains field windings.

step-down transformer: device that reduces voltage (e.g., between primary distribution and secondary distribution).

step potential: voltage differential that develops when a person near a ground fault with two parts of their body (two feet, a hand and a foot or another body part) straddles that voltage differential. It can result in electrical shock or electrocution.

step-up transformer: device that increases voltage (e.g., from distribution to transmission).

storm center: facility used by utilities and contractors to monitor storms and direct storm responses.

storm drill: to fix something in the mind or habit pattern by repetitive instruction, in this case, storm response procedures.

storm recovery: process of repairing damage and restoring services following a storm.

storm response: the collective efforts of governments, utilities, contractors, first responders, and volunteers to bring relief to areas stricken by storms. In utility arboriculture, the combined efforts of utilities and contractors to mobilize and assist in the restoration of utility services.

storm surge: rise and onshore flow of seawater as the result of the winds of a storm, and also the surface pressure drop near the storm center.

straight-line winds: powerful winds generally greater than 50 mph (80 km/h) associated with outflow from thunderstorms or derechos, and not associated with tornadoes.

stress: action on a body of any system of balanced forces whereby strain or deformation results.

strike zone: the area where trees are located that are sufficiently tall to contact an overhead conductor should they fall.

stump application: individual stem treatment in which herbicides are applied to the stump surface around the cambium and topside of the bark.

subordination: pruning that reduces the size and ensuing growth of a branch in relation to other branches or leaders.

substation: electric facility equipped with transformers, switching equipment, and protection and control devices such as circuit breakers, automatic line reclosers, capacitors, or voltage regulators. Voltage is changed in substations.

subtransmission lines: high-voltage lines generally energized between 69 and 161 kV. They can be as low as 35 kV. Subtransmission lines connect bulk transmission substations to industrial customers or distribution substations.

supervisory control and data acquisition (SCADA): system that automatically collects data and enables remote control switching operations.

supply container: container in which custom blends of diluted concentrates, or ready-to-apply herbicides, are provided to applicators. Registered herbicide concentrate is typically provided in returnable, reusable supply containers, while registered and ready-to-use products are typically provided in one-way, disposable supply containers.

switchyard: electric facility that does not change voltage, but is used to route power through various circuits. May be strategically designed to compensate for portions of a system that are experiencing power failures, protecting circuits through disconnect switches, circuit breakers, relays, and communications systems (compare to *substation*).

SWOT: in project management, analysis of strengths, weaknesses, opportunities, and threats.

synthetic auxin: class of selective herbicides that kills broadleaved plants, but does not harm grasses. Artificial form of auxin, a natural plant growth regulator.

T

taper: change in diameter over the length of trunks, branches, and roots.

taps: wire connection between energized conductors and other electrical equipment, such as transformers.

targeted messaging: using various media to deliver specific information to individual customers or groups of customers.

teratogenicity: ability of a substance to produce birth defects.

terminal bud: bud at the tip of a twig or shoot. Apical bud (contrast with *lateral bud*).

thunderstorm: local storm produced by a cumulonimbus cloud and accompanied by lightning and thunder, with strong gusts of wind, heavy rain, and sometimes hail.

time and material contract: method of providing a service in which all costs—including labor, equipment, supplies, overheads, and operating profit—are passed on to the contracting agency, usually on an hourly basis.

tolerance level: maximum incompatible plant pressures (species, density, height, location, or condition) allowable before unacceptable consequences develop.

topiary: pruning system that uses a combination of pruning, supporting, and training branches to orient a plant into a desired shape.

topping: reducing tree size by cutting live branches and leaders to stubs, without regard to long-term tree health or structural integrity.

tornado: rotating column of air, in contact with the earth's surface, pendant from a cumuliform cloud, often visible as circulating debris/dust at the ground.

touch potential: voltage differential between two objects someone simultaneously contacts.

toxicity: ability of a substance to damage an organ system, to disrupt a biochemical process, or disturb an enzyme system.

TPZ: see *tree protection zone*.

track forecast cone: the probable track of the center of a tropical cyclone based on forecasts from the previous 5 years, with an accuracy rate of approximately 67 percent.

transformer: electrical device that raises or lowers voltage through induction.

transient fault: fault that affects the dielectric properties of a system for an instant, and no longer exists after the power has been restored.

transmission interconnect: region where transmission or extra-high transmission lines are connected. The interconnection allows utilities to transfer electricity both within their systems and among one another.

transmission lines: in an electric utility system, wires used to transmit electricity from generating stations to the distribution network or between distribution substations, often carrying in excess of 69,000 volts (contrast with *distribution lines*).

tree growth regulator (TGR): chemical that slows terminal growth by reducing cell elongation.

tree protection zone (TPZ): defined area within which certain activities are prohibited or restricted to prevent or minimize potential injury to designated trees, especially during construction or development.

tree risk assessment: a systematic process used to identify, analyze, and evaluate tree risk.

tree wire: coated overhead primary used to provide some protection from faults caused by branches. It is not insulation and is no safer than uncovered wire.

triple constraint: in project management, a concept showing the interrelation of the resources of time, cost, and scope, often depicted in a constraint triangle.

triplex: secondary wire, which has two insulated conductors and a neutral wrapped around a supportive cable. In North America, a single conductor typically carries 120 V, and the two together total 240 V.

tropical cyclone: cyclone that originates over the tropical oceans, including tropical depressions, tropical storms, hurricanes, and typhoons.

trunk injection: technique to introduce substances directly into the xylem of a tree to treat or prevent diseases, disorders, or pest problems.

turn-key contract: contract arrangement in which the contractor is responsible for all financing, and owns the work until the project is complete and turned over to the owner.

turns ratio: ratio between the two wire coils in a transformer; identical to voltage ratio.

typhoon: regional term (Western North Pacific) for a severe tropical cyclone.

U

underground construction: belowground utility installation.

unit price contract: method of providing a service or product where an agreed-upon amount is paid for each designated amount, or unit, of work completed. In vegetation management, units can be trees, portions of trees, miles of line, spans, or others as agreed by all parties involved.

unsafe acts: work practices such as willful rule violation and maliciousness, as well as human error, which could lead to injury or a fatality.

unsafe working conditions: working circumstances, such as equipment failure, that could lead to injury or fatality.

urban forest: the sum of all woody and associated vegetation in and around dense human settlements, from small communities to metropolitan regions, including street, residential, and park trees, greenbelt vegetation, trees on unused public and private land, trees in transportation and utility corridors, and forests on watershed lands.

urban forestry: management of naturally occurring and planted trees and associated plants in urban areas.

utility facilities: infrastructure used to provide utility services, such as poles, wires, transformers, switches, etc.

utility forest: the population of trees that could now or in the future interfere with the operation of utility facilities.

utility line-clearance pruning: selective removal of vegetation, especially tree branches, which could affect electric supply lines or other utility facilities.

utility pruning: pruning around or near utility facilities with the object of maintaining safe and reliable utility service.

utility service: a product or service provided by a utility company, such as electricity, telecommunications, or natural gas.

utility vegetation manager: a professional with the proper experience, education, and training to successfully establish or supervise an integrated vegetation management program.

vehicle monitoring system (VMS): GPS-based system capable of providing vehicle location and other data in real time, as well as movement history.

vigor: overall health; capacity to grow and resist stress.

VMS: see *vehicle monitoring system.*

volt (V): measure of electric force or pressure, also called *electromotive force.* Volts are analogous to pounds per square inch or kilopascals in a hydraulic system.

voltage gradient: electrical potential over a specified distance. In the context of tree-caused electrical faults, it is determined by the voltage and spacing of the lines, as well as stem diameter and species of tree. Greater branch diameter and closer phase spacing create higher gradients. The higher the gradient, the more likely a tree is to cause a fault.

voltage ratio: the proportion of primary to secondary voltage of a transformer. It is identical to the turns ratio such that if the primary winding has ten times the number of turns as the secondary coil, the turns ratio is 10 to 1; the voltage will be stepped down by a factor of 10.

voltage regulator: electrical device that functions like an adjustable transformer, capable of either increasing or decreasing the circuit voltage.

warning (storm): statement issued by a government weather agency indicating an imminent threat of severe or hazardous weather.

watch (storm): statement issued by a government weather agency indicating a potential threat of severe or hazardous weather.

water head: pressure built up behind hydroelectric dams and used to drive turbines to generate electricity.

watt (W): a measure of electric power; one amp flowing at one volt. Calculated by multiplying volts and amps.

Weingarten rights: workers' rights established by the United States Supreme Court in the 1975 case of *National Labor Relations Board v. J. Weingarten, Inc.* The judgment guarantees an employee the right to Union representation during an investigatory interview.

whorl: a pattern of leaves, twigs, or branches arranged in a circle around a point on the stem.

windrow: a line of cut and piled brush.

wire-border zone: transmission right-of-way vegetation management philosophy applied through cover-type conversion. The wire zone is the section of a utility transmission right-of-way under the wires and extending out both sides to a specified distance. The border zone is the remainder of the right-of-way, where small trees and tall shrubs (under 25 feet [8 m] in height at maturity) are established.

wire zone: see *wire-border zone.*

work breakdown structure: project management stratification of the entire scope of work needed to accomplish a project's objectives.

workload assessment: survey of the volume of work. May be done by comprehensive inventories or sampling.

wound treatment: material applied to pruning cuts or other openings in the bark of trees, to prevent the spread of pests, or for other specified reasons.

Wye construction: distribution construction consisting of three phases and a grounded or neutral wire. The phases have polarity, creating a voltage differential among them.

REFERENCES

Abbott, R.E., P.J. Dubish, and J.T. Rooney. 2005. *Line Clearance Arboriculture: Field Guide to Technique, Efficiency and OSHA/ANSI Compliance.* Cuyahoga Falls, Ohio: ACRT, Inc.

Abrashoff, M.D. 2002. *It's Your Ship: Management Techniques from the Best Damn Ship in the Navy.* New York: Warner Books, Inc.

AccuWeather Glossary. *What is a snow ratio?* Retrieved January 9, 2017 from http://www.accuweather.com/en/weather-glossary/what-are-snow-ratios/4789125

Agnew, J., and A. Daniels. 2010. *Safe by Accident? Take the Luck out of Safety: Leadership Practices that Build a Sustainable Safety Culture.* Atlanta, GA: Performance Management Publications.

American Meteorological Society. 2012. *Glossary of meteorology.* Retrieved January 9, 2017 from http://glossary.ametsoc.org/wiki/Cyclones

American National Standards Institute (ANSI). 2008. *American National Standard for Tree Care Operations: Tree, Shrub, and Other Woody Plant Management – Standard Practices (Pruning).* (A300 Part 1). Londonderry, New Hampshire: Tree Care Industry Association.

American National Standards Institute (ANSI). 2012. *American National Standard for Tree Care Operations: Tree, Shrub, and Other Woody Plant Management – Standard Practices (Integrated Vegetation Management a. Utility Rights-of-Way)* (A300 Part 7). Londonderry, New Hampshire: Tree Care Industry Association.

American National Standards Institute (ANSI). 2017a. *American National Standard for Arboricultural Operations – Safety Requirements* (Z133). Champaign, Illinois: International Society of Arboriculture.

American National Standards Institute (ANSI). 2017b. *American National Standard for Tree Care Operations: Tree, Shrub, and Other Woody Plant Management – Standard Practices (Pruning).* (A300 Part 1). Londonderry, New Hampshire: Tree Care Industry Association.

American National Standards Institute (ANSI). 2017c. *American National Standard for Tree Care Operations: Tree, Shrub, and Other Woody Plant Management – Standard Practices (Tree Risk Assessment a. Tree Failure)* (A300 Part 9). Londonderry, New Hampshire: Tree Care Industry Association.

American National Standards Institute (ANSI). 2017d. *National Electrical Safety Code (ANSI C2).* Washington, D.C.: American National Standards Institute.

Amo, T. *The negative effects of a lack of training in the workplace.* Retrieved January 10, 2017 from http://smallbusiness.chron.com/negative-effects-lack-training-workplace-45171.html

Anderson, L.M., and T.A. Eaton. 1986. Liability for damage caused by hazardous trees. *Journal of Arboriculture,* 12:189–195.

Appelt, P. and D. Gartman. 2004. *Integrated vegetation management on natural gas pipeline rights-of-way.* Presentation to the Environmental Concerns in Rights-of-Way Management 8th International Symposium. Saratoga Springs, NY. September 12–16, 2004.

Ardichvili, A.A., D.J. Jondle, and J.A. Mitchell. 2009. Characteristics of ethical business cultures. *Journal of Business Ethics,* 85:445–451.

Australian Government Bureau of Meteorology. n.d. (a) *About tropical cyclones.* Retrieved January 9, 2017 from http://www.bom.gov.au/cyclone/about/

Australian Government Bureau of Meteorology. n.d. (b) *Tornado, twister hurricane, tropical cyclone, typhoon – what's the difference?* Retrieved January 9, 2017 from http://media.bom.gov.au/social/blog/6/tornado-twister-hurricane-tropical-cyclone-typhoon-whats-the-difference/

Baca, C.M. 2007. *Project Management for Mere Mortals*. Boston: Addison-Wesley.

Bai, S., W.R. Chaney, and Y. Qi. 2004. Response of cambial and shoot growth in trees treated with paclobutrazol. *Journal of Arboriculture,* 30:137–145.

Bai, S., W.R. Chaney, and Q. Yadong. 2005. Wound closure in trees affected by paclobutrazol. *Journal of Arboriculture,* 31:273–279.

Ball, D., P.S. Coelho, and A. Machás. 2004. The role of communication and trust in explaining customer loyalty: An extension to the ECSI model. *European Journal of Marketing,* 38:1272–1293.

Ball, J. 2014. Tree worker incidents; it's déjà vu all over again. *Tree Care Industry,* XXV(1):8–10.

Ball, J., and M. Johnson. 2015. First aid part 5: work related trauma: Injuries overview. *Tree Care Industry,* XXVI(5):38–40.

Ball, J., and S. Vosberg. 2010. A survey of United States tree care companies: Part 1 - safety training and fatal accidents. *Arboriculture & Urban Forestry,* 36:224–229.

Bayer AG. Krenite® S Specimen Label. 2018. Retrieved March 29, 2018 from https://www.backedbybayer.com/~/media/BackedByBayer/Product%20Labels%20-%20pdf/Krenite%20S.ashx

Bedard, R. 2007. *Power and Energy from the Ocean Energy Waves and Tides: A Primer*. Washington, D.C.: Electric Power Research Institute.

Bell, R. 2004. Bidding out and bidding on a unit price contract. *Utility Arborist Quarterly,* Fall 2004:2–4.

Bennis, W. 2009. *On Becoming a Leader*. New York: Basic Books.

Berry T., and D. Wilson. 2005. *On Target: The Book on Marketing Plans.* Eugene, Oregon: Palo Alto Software, Inc.

Blair, D.F. 1989. Safety training for the professional and the non-professional. *Journal of Arboriculture,* 15:209–214.

Blair, D.F. 1995. *Arborist Equipment: A Guide to the Tools and Equipment of Tree Maintenance and Removal*. Champaign, Illinois: International Society of Arboriculture.

Blair, G.D. 1940. *Tree Clearance for Overhead Lines: A Textbook of Public Utility Forestry*. Chicago, Illinois: Electrical Publications, Inc.

Botanical Society of Edinburgh. 1873. *Transactions,* Volume 11, Issue 1–4, p. 453.

Bramble, W.C., Yahner, R.H., Byrnes, W.R., and S.A. Liscinsky. 1992. Small mammals in plant cover types on an electric transmission right-of-way. *Journal of Arboriculture,* 12:316–321.

Brennan, L.A. 2012. *Comparison and assessment of mechanical and herbicide-chemical side-trimming methods of managing roadside vegetation by the Texas Department of Transportation (TxDOT)*. Technical Report 0-6732-1. Retrieved January 11, 2017 from Texas A&M University-Kingsville Web site: http://tti.tamu.edu/documents/0-6732-1.pdf

British Standards Institute. 2012. Trees in relation to design, demolition and construction - Recommendations. *BS,* 5837:2012.

Browning, D.M., and H.V. Wiant. 1997. The economic impacts of deferring electric utility tree maintenance. *Journal of Arboriculture,* 23:106–112.

California Public Resources Code. Division 4 (Forests, Forestry and Range and Forage Lands), Part 2 (Protection of Forest, Range and Forage Lands), Chapter 3 (Mountainous, Forest-, Brush- and Grass-Covered Lands), Section 4293. Retrieved January 6, 2017 from http://leginfo.legislature.ca.gov/faces/codes_displaySection.xhtml?lawCode=PRC&sectionNum=4293

California Public Utilities Commission. General Order 95, Section III, Requirements for All Lines, Rule 35, Vegetation Management. Sacramento, California: California Public Utilities Commission. Retrieved January 9, 2017 from http://www.cpuc.ca.gov/gos/GO95/go_95_rule_35.html

Campbell, R.J. 2012. *Weather-related power outages and electric system resiliency.* Congressional Research Service Report for Congress 7-5700. Retrieved January 6, 2017 from http://fas.org/sgp/crs/misc/R42696.pdf

Carson, R. 1962. *Silent Spring.* New York: Crest Books.

Cheney, W.R. 2005. *Growth Retardants: A Promising Tool for Managing Urban Trees.* Purdue Extension, FNR-252-W.

Chick, T. 2010. Allopathy as a biological control for integrated vegetation management. *Arborist News,* 19(6):18–20.

Chisholm, M. n.d. *Tips for tree care and storms: Before, during and after.* Retrieved March 1, 2018 from https://www.stihlusa.com/information/articles/tips-for-tree-care-and-storms/

Cieslewicz, S.R., and W.H. Porter. 2010. *CN Utility Consulting Utility Vegetation Management Benchmark and Industry Intelligence.* Sebastopol, California: CN Utility Consulting, Inc.

Cieslewicz, S.R., and R.R. Novembri. 2004. [Report prepared by CN Utility Consulting, LLC] *Utility Vegetation Management Final Report: Commissioned to Support the Federal Investigation of the August 14, 2003 Northeast Blackout.* Washington, D.C.: Federal Energy Regulatory Commission.

Clapp, A.L. 2006. *NESC Handbook: A Discussion of the National Electric Safety Code®.* New York: Standards Information Network IEEE Press.

Clapp, A.L. 2007. *National Electrical Safety Code Handbook.* (6th Ed.). New York: Institute of Electrical and Electronics Engineers.

Clark, R.W., J. Halligan, E. Moldawsky, and R. Stepanian. 2008. *Report of the Consumer Protection and Safety Division Regarding the Guejito, Witch and Rice Fires.* Petition 7-11-007 before the Public Utilities Commission of the State of California.

Cline, M.G. 1997. Concepts and terminology of apical dominance. *American Journal of Botany,* 84:1064–1069.

CNUC. 2010a. *CN Utility Consulting Utility Vegetation Management Benchmark and Industry Intelligence.* Des Moines, Iowa: CN Utility Consulting.

CNUC. 2010b. *Regulatory Requirements 2006–2009: Assessment Performed for the 2010 CN Utility Consulting Utility Vegetation Management Benchmark & Industry Intelligence.* Sebastopol, California: CN Utility Consulting. Retrieved January 16, 2017 from http://www.cnutility.com/wp-content/uploads/2013/10/UVM-Regulatory-Requirements-2006-2009.pdf

CNUC. 2015. *Distribution Utility Vegetation Management Benchmark & Intelligence 2014 Distribution Update.* Chicago, Illinois: CN Utility Consulting.

Commissioners fire GM of Lansing-owned utility. 2015, January 14. *Detroit News.* Retrieved January 9, 2017 from http://www.dailyherald.com/article/20150114/news/301149958/

Commonwealth of Massachusetts, General Laws, Title, XIV, Public Ways and Works, Chapter 87, Shade Trees, Section 14. Retrieved January 13, 2017 from https://malegislature.gov/Laws/GeneralLaws/PartI/TitleXIV/Chapter87/Section14

Corfidi, S., J.S. Evans, and R.H. Johns. n.d. *About derechos.* Retrieved January 9, 2017 from http://www.spc.noaa.gov/misc/AbtDerechos/derechofacts.htm

Critical infrastructure assurance guidelines for municipal governments: Planning for electric power disruptions. 2001. Retrieved January 10, 2017 from http://www.osti.gov/scitech/servlets/purl/807351

Crompton, J.L. 1999. *Financing and Acquiring Park and Recreation Resources.* Champaign, Illinois: Human Kinetics.

Curtis, J.T. 1971. *The Vegetation of Wisconsin: An Ordination of Plant Communities.* Madison: The University of Wisconsin Press.

Dahle, G.A., H.H. Holt, W.R. Chaney, T.M. Whalen, D.L. Cassens, R. Gazo, and R.L. McKenzie. 2006 Branch strength loss implications for silver maple (*Acer saccharinum*) converted from round-over to V-trim. *Arboriculture & Urban Forestry*, 32:148–154.

DelPo, A. 2007. 2nd Ed. *The Performance Appraisal Handbook: Legal and Practical Rules for Managers.* Berkeley, CA, p. 197.

Deric, M., and R. Hollenbaugh. 2003 (March). Vegetation managers weigh costs against performance. *Transmission & Distribution World*, 55.

Dolce, J. 1984. *Fleet Management.* New York: McGraw Hill.

Dolce, J.E. 1994. *Analytical Fleet Maintenance Management.* Warrendale, Pennsylvania: Society of Automotive Engineers.

Dorf, R.C. (Ed.). 1993. *The Electrical Engineering Handbook.* Ann Arbor, Michigan: CRC Press.

Draper, J. 1892. *An essay on "The preservation of roadside trees and the improvement of public grounds."* Publications of the Massachusetts Society for Promoting Agriculture. Retrieved January 11, 2017 from http://www.archive.org/details/essayonthepreser00drap

Driggs, K. 2005, October. *Overview of the power grid in the west.* Paper presented at the Promoting Effective Collaborations Conference of the Western Electricity Coordinating Council, Salt Lake City, Utah.

Drucker, P.F. 2001. *The Essential Drucker: The Best of Sixty Years of Peter Drucker's Essential Writings on Management.* New York: Harper Collins.

Duffy, J. *The best project management software of 2017. PCmag.com* Retrieved January 9, 2017 from http://www.pcmag.com/article2/0,2817,2380448,00.asp

Dunster, J.A., E.T. Smiley, N. Matheny, and S. Lilly. 2017. *Tree Risk Assessment Manual.* (2nd Ed.). Champaign, Illinois: International Society of Arboriculture.

Dweck, C.S. 2006. *MINDSET: The New Psychology of Success.* New York: Random House.

Edison Electric Institute, 2014. *Before and After the Storm: A Compilation of Recent Studies, Programs, and Polices Related to Storm Hardening and Resiliency.* Washington, D.C.: Edison Electric Institute.

Edwards, R. 2016. *The online tornado FAQ.* NOAA Storm Prediction Center. Retrieved January 6, 2017 from www.spc.noaa.gov/faq/tornado

Electric Light and Power. 2006 (May 1). *Managing vegetation: With the public in mind.* Retrieved March 1, 2018 from http://www.elp.com/articles/print/volume-84/issue-3/transmission-distribution/managing-vegetation-with-the-public-in-mind.html

Energy Networks Association, 2007. *Vegetation Management Near Electricity Equipment- Principles of Good Practice* (ENA ET ETR 136-2007). London, U.K.: Energy Networks Association Publications.

Electric Light and Power. 2014 (May 12). *Penelec to conduct storm outage restoration drill.* Retrieved March 1, 2018 from http://www.elp.com/articles/2014/05/penelec-to-conduct-storm-outage-restoration-drill.html

Energy Policy Act of 2005, Pub. L. No. 109-58, Section 1211. 2005. Retrieved January 10, 2017 from https://www.gpo.gov/fdsys/pkg/PLAW-109publ58/pdf/PLAW-109publ58.pdf

European Commission Directorate-General for Energy and Transport. 2006. *Study on the technical security rules of the European Electricity Network*. Final Report 62236A001 Rev. 2. Retrieved January 9, 2017 from https://ec.europa.eu/energy/sites/ener/files/documents/2006_02_security_rules_pb_power.pdf

Evans, R. 2007. *Fueling Our Future: An Introduction to Sustainable Energy*. Cambridge, U.K.: Cambridge University Press.

Fazio, J.R. 1992. *Trenching & Tunneling Near Trees: A Field Pocket Guide for Qualified Utility Workers*. Nebraska City, Nebraska: The National Arbor Day Foundation.

Federal Emergency Management Agency, Incident Command System (ICS) Resource Center. n.d. *Glossary of related terms*. Retrieved January 6, 2017 from https://training.fema.gov/emiweb/is/icsresource/glossary.htm

Federal Emergency Management Agency. n.d. *The disaster declaration process*. Retrieved January 6, 2017 from https://www.fema.gov/declaration-process-fact-sheet

Federal Emergency Management Agency. 2013. *Hurricane Sandy FEMA after-action report*. Retrieved January 9, 2017 from https://www.fema.gov/media-library-data/20130726-1923-25045-7442/sandy_fema_aar.pdf

FirstEnergy. n.d. *Communication tools*. Retrieved January 6, 2017 from https://www.firstenergycorp.com/content/customer/help/communication_tools.html

Fitzpatrick, M., and M. Fitzpatrick. 2014, August 20. Ditching our distractions: The importance of active listening. Retrieved January 6, 2017 from http://www.forbes.com/sites/sungardas/2014/08/20/ditching-our-distractions-the-importance-of-active-listening/

Florida Public Service Commission. 2006. Notice of proposed agency action order requiring storm implementation plans. Order No. PSC-06-035 1-PAA-EI. Retrieved January 6, 2017 from http://www.psc.state.fl.us/library/filings/06/03645-06/03645-06.PDF

Fortin Consulting, Inc.; University of Minnesota Sea Grant Program; and the Natural Resources Research Institute, University of Minnesota Duluth. 2014. *Field guide for maintaining rural roadside ditches*. Retrieved January 11, 2017 from http://www.lakesuperiorstreams.org/stormwater/toolkit/contractor/resources/DitchGuide_SeaGrant.pdf

French, S.C., and B.L. Appleton. 2009. *A guide to successful pruning: Stop topping trees!* Virginia Cooperative Extension Publication 430-458. Retrieved January 11, 2017 from Virginia Tech Web site: http://pubs.ext.vt.edu/430/430-458/430-458.html

Gaskill, A. 1918. *A shade tree guide*. Reports of the Department of Conservation and Development, State of New Jersey. Union Hill, New Jersey: Hudson Printing Company. Retrieved January 9, 2017 from https://archive.org/details/cu31924002856502

Gebelein, S.H., B. Davis, K.J. Nelson-Neuhaus, and C.J. Skube. 2004. *Successful Manager's Handbook*. (7th Ed.). Minneapolis, Minnesota: Personnel Decisions International.

Gilman, E.F., B. Kempf, N. Matheny, and J. Clark. 2013. *Structural Pruning: A Guide for the Green Industry*. Visalia, California: Urban Tree Foundation.

Gipe, P. 1995. *Wind Energy Comes of Age*. New York: John Wiley & Sons.

Glasstone, S., and W.H. Jordon. 1980. *Nuclear Power and its Environmental Effects*. LaGrange Park, Illinois: American Nuclear Society.

Gober, D.L. 2012. Struck-by incidents. *Utility Arborist Newsline*, 3:15–16.

Goleman, D.R., R.E. Boyatzis, and A. McKee. 2002. *Primal Leadership: Learning to Lead with Emotional Intelligence.* Boston: Harvard Business School Press.

Goodfellow, J.W. 1985. Fixed price bidding of distribution line clearance work: Another look. *Journal of Arboriculture,* 11:116–120.

Goodfellow, J.W. 1995. Engineering and construction alternatives to line clearance tree work. *Journal of Arboriculture,* 21:41–49.

Goodfellow, J.W. 2005. Investigating tree-caused faults. *Transmission & Distribution World,* 57 (November):8–10.

Goodfellow, J.W. 2008. Report to ConEd New York Public Service Commission (December 28, 2008).

Goodfellow, J.W. 2010. Research, development, & demonstration in utility arboriculture. *Arborist News,* 19(February):48–51.

Goodfellow, J.W., B. Blumreich, and G. Nowacki. 1987. Tree growth response to line clearance pruning. *Journal of Arboriculture,* 13:196–200.

Goodfellow, J.W., A. Detter, P. van Wassenaer and M. Neiheimer. 2009. *Final Report: Development of Risk Assessment Criteria for Branch Failures Within the Crowns of Trees.* BioCompliance Consulting, Inc.

Goodfellow, J.W., and H.A. Holt. 2011. *Utility Arborist Association Best Management Practices Field Guide to Closed Chain of Custody for Herbicide Use in the Utility Vegetation Management Industry.* Champaign, Illinois: International Society of Arboriculture.

Gore, Al. 2000. *Earth in the Balance.* Boston: Houghton Mifflin.

Government of Newfoundland and Labrador, Canada, Department of Advanced Education, Skills and Labour. n.d. *Grievance mediation.* Retrieved January 12, 2017 from http://www.aesl.gov.nl.ca/labour/union/grievancemediation.html

Graham, C.B. amd S.W. Hays. 1986. *Managing the Public Organization.* Washington, D.C.: CQ Press.

Graham, S.A. 1945. Ecological classifications of cover types. *Journal of Wildlife Management,* 9:182–190.

Grayson, L. n.d. *UAA best management practices: Funding* (Utility Arborist Association White Paper). Retrieved March 29, 2018 from http://gotouaa.org/wp-content/uploads/2018/03/fundingWP.pdf

Guggenmoos, S. 2003. Effects of tree mortality on power line security. *Journal of Arboriculture,* 29:181–196.

Guggenmoos, S. 2011 (March 7). Vegetation management terms. *T&D World Magazine.* Retrieved March 29, 2018 from http://www.tdworld.com/insights/vegetation-management-terms

Hager, A.C., and D. Refsell. 2008. Herbicide persistence and how to test for residues in soils. In *Illinois Agricultural Pest Management Handbook* (pp. 279–286). Champaign, Illinois: University of Illinois Extension.

Harris, R.W., J.R. Clark, and N.P. Matheny. 2004. *Arboriculture: Integrated Management of Landscape Trees, Shrubs, and Vines* (4th Ed.). Upper Saddle River, New Jersey: Prentice Hall.

Hashemian, H.M., and W.C. Bean. 2011 (October). State-of-the-art predictive maintenance techniques. *IEEE Transactions on Instrumentation and Measurement,* 60(10):3480–3492.

Heinrich, H.W., D.C. Petersen, and N.R. Roos. 1980. *Industrial Accident Prevention* (5th Ed.). New York: McGraw-Hill Book Company.

Herda, D.J., and M.M. Madden. 1991. *Energy Resources Towards a Renewable Future.* New York: Science and Technology Society.

Hightshoe, G.L. 1988. *Native Trees, Shrubs and Vines for Urban and Rural America: A Plant Design Manual for Environmental Engineers.* New York: Van Norstrand Reinhold.

Hodel, D.R. 2009. Biology of palms and implications for management in the landscape. *HortTechnology*, 19:676–681.

Homer, C.G., J.A. Dewitz, L. Yang, S. Jin, P. Dalialson, G. Xian, J. Colston, N.D. Harold, J.D. Wickhan, and K. Megown. 2015. Completion of the 2011 National Land Cover Database for the conterminous United States—Representing a decade of land cover change information. *Photogrammetric Engineering and Remote Sensing*, 81(5):345–354.

Hooper, B. 2003. Vegetation management takes to the air. *Transmission & Distribution World*, 55 (September).

Horn, J.W., E.B. Arnett, and T.H. Kunz. 2006. Behavioral responses of bats to operating wind turbines. *Journal of Wildlife Management*, 72:123–132.

Hudson, B. 1983. Private sector business analogies applied in urban forestry. *Journal of Arboriculture*, 9:253–258.

Indiana Regulatory Commission. 2012. *Tree trimming and vegetation management*. Retrieved March 7, 2018 from https://www.in.gov/oucc/2555.htm

Institute of Electrical and Electronic Engineers. 2012. *National Electrical Safety Code (NESC)* (ANSI C2). New York: Institute of Electrical and Electronics Engineers, Inc.

Institute of Electrical and Electronic Engineers; International Electrotechnical Commission. 1983. *IEC Multilingual Dictionary of Electricity*. New York: Institute of Electrical and Electronics Engineers in cooperation with the International Electrotechnical Commission.

International Brotherhood of Electrical Workers and National Electrical Contractors Association. 2006. *Underground Distribution: How to Install Cables, Fuses, Transformers, Grounds, and Switches on Underground Systems* (2nd Ed.). International Brotherhood of Electrical Workers and National Electrical Contractors Association in conjunction with the National Joint Apprenticeship and Training Committee. Newport Beach, California: Alexander Publications.

International Code Council. 2006. *Urban-Wildland Interface Code*. Washington, D.C.: International Code Council.

International Society of Arboriculture. 2015. *Glossary of Arboricultural Terms*. Champaign, Illinois: International Society of Arboriculture.

International Society of Arboriculture. n.d. *ISA Certified Arborist Code of Ethics*. Retrieved March 7, 2018 from http://www.isa-arbor.com/Portals/0/Assets/PDF/Certification/Ethics-Code-of-Ethics.pdf

Jacobs, P. 2012. Achieving compliance with the law and company rules. *Utility Arborist Newsline*, 3:30.

James, K. 2003. Dynamic loading of trees. *Journal of Arboriculture*, 29:165–171.

J.D. Power. 2014, July 16. *Overall customer satisfaction with residential electric utilities continues to improve; however, utilities not keeping pace with satisfaction increases in other service industries*. Retrieved January 12, 2017 from http://www.jdpower.com/press-releases/2014-electric-utility-residential-customer-satisfaction-study#sthash.cbR0E1ff.dpuf

Johns, H.R., and D.E. Holewinski. 1981. How to develop or improve your line clearing program. *Journal of Arboriculture*, 7:329–332.

Johnson, D., and C. Pedersen. 2014 (June 16). *A Utility's Social Media Journey to Customer Connectivity*. Electric Light and Power. Retrieved January 31, 2018 from http://www.elp.com/articles/print/volume-92/issue-3/sections/it-cis-crm/a-utility-s-social-media-journey-to-customer-connectivity.html

Johnson, P.M. 2005. *A glossary of political economy terms*. Retrieved January 6, 2017 from Auburn University, Department of Political Science Web site: http://www.auburn.edu/~johnspm/gloss/

Johnstone, R.A., and M.R. Haggie. 2012. Regional vegetation management best practices case studies: an applied approach for utility and wildlife managers. In J.M. Evans, J.W. Goodrich-Mahoney, D. Mutrie, and J. Reinemann (Eds.), *Environmental Concerns in Rights-of-Way Management 9th International Symposium* (pp. 77–86). Champaign, Illinois: International Society of Arboriculture.

Kempter, G.P. 2004. *Best Management Practices: Utility Pruning of Trees.* Champaign, Illinois: International Society of Arboriculture.

Kendrick, T. 2004. *The Project Management Tool Kit: 1001 Tips and Techniques for Getting the Job Done Right.* New York: American Management Association.

Kimball, S.L. 1990. The physiology of tree growth regulators. *Journal of Arboriculture,* 16:39–41.

Kligman, D. (2012, September 13). PG&E's tree trimming protects electric lines, reduces outages. *PG&E Currents.* Retrieved January 9, 2017 from http://www.pgecurrents.com/ 2012/09/13/ pge%E2%80%99s-tree-trimming-protects-electric-lines-reduces-outages/

Krause, D. 2018. *Role of Herbicides in IVM.* Presentation to Michigan UAA Regional Meeting. Lansing, MI. February 12, 2018.

Kuhns, M.R., and D.K. Reiter. 2007. Knowledge of and attitudes about utility pruning and how education can help. *Arboriculture & Urban Forestry,* 33:264–274.

Larson, E. 2000. *Isaac's Storm: A Man, a Time, and the Deadliest Hurricane in History.* New York: Vintage Books.

Library of Congress. n.d. Today in history – May 24. *What hath God wrought?* Retrieved January 13, 2017 from https://www.loc.gov/item/today-in-history/may-24

Liu, D., T.R. Mitchell, T.W. Lee, B.C. Holtom, and T.R. Hinkin. 2012. When employees are out of step with coworkers: How job satisfaction trajectory and dispersion influence individual- and unit-level voluntary turnover. *Academy of Management Journal,* 55:1360–1380.

Lloyd, J., and F. Miller. 1997. Introduction to plant health care. In *Plant Health Care for Woody Ornamentals* (pp. 1–4). Savoy, Illinois: International Society of Arboriculture.

Local Government Association of South Australia. 2014. *Directions for vegetation management: SA Power Networks long-term plan for managing trees near powerlines.* Unpublished discussion paper for the Vegetation Management Working Group, Local Government Association board meeting (19 June 2014). Retrieved January 13, 2017 from https://www.aer.gov.au/

Lough, W.B. 1991. Contracting for urban tree maintenance. *Journal of Arboriculture,* 17:16–17.

Luefschuetz, G.S. 2010. *Selling Professional Services to the Fortune 500: How to Win in the Billion-Dollar Market of Strategy Consulting, Technology Solutions and Outsourcing Services.* New York: McGraw Hill.

Maine Municipal Association, Risk Management Services, Loss Control. 2011. *Best practice: Safety committees.* Retrieved January 11, 2017 from https://ehs.utexas.edu/programs/safety/documents/SafetyCommittee-BestPractices.pdf.

Mann, M.P., H.A. Holt, W.R. Chaney, W.C. Mills, and R.L. McKenzie. 1995. Tree growth regulators reduce line clearance trimming time. *Journal of Arboriculture,* 21:209–212.

Matheny, N.P., and J.R. Clark. 1994. *A Photographic Guide to the Evaluation of Hazard Trees in Urban Areas* (2nd Ed.). Urbana, Illinois: International Society of Arboriculture.

Matheny, N.P., and J.R. Clark. 1998. *Trees and Development: A Technical Guide to Preservation of Trees During Land Development.* Champaign, Illinois: International Society of Arboriculture.

Matheny, N.P., and J.R. Clark. 2008. *Municipal Specialist Certification Study Guide.* Champaign, Illinois: International Society of Arboriculture.

McCabe J., and A. Koeser. (2010, September). Western Integrated Management Summit Research Summit White Paper. Forest Lake, Minnesota: Utility Arborist Association.

McClenahan, J. 2012. Risk management: Utilization of leading indicators in the continuous improvement cycle. *Tree Care Industry*, XXIII(11):68-72.

McConnell, D.W., R.L. Mahoney, W.M. Colt, and A.D. Partridge. 1998. *How to prune coniferous evergreen trees*. University of Idaho, College of Agriculture, Cooperative Extension Bulletin No. 644. Retrieved January 11, 2017 from http://www.cals.uidaho.edu/edcomm/pdf/bul/bul0644.pdf

McManus, M., N. Schneeberger, R. Reardon, and G. Mason. 1989. *Gypsy moth*. Forest Insect & Disease Leaflet 162. U.S. Department of Agriculture, Forest Service.

McPherson, E.G. 2003. A benefit-cost analysis of ten street tree species in Modesto, California, U.S. *Journal of Arboriculture*, 29:1–8.

Meehan, B. 2007. *Empowering Electric and Gas Utilities with GIS*. Redlands, California: ESRI Press.

Mendelsohn, M., T. Lowder and B. Canavan. 2012. *Utility-Scale Concentrating Solar Power and Photovoltaics Projects: A Technology and Market Overview*. Technical Report NREL/TP-6A20-51137. National Renewable Energy Laboratory, Golden Colorado.

Miller, R.H. 2002. Danger, high voltage! *Tree Care Industry*, XIII(1):9–15.

Miller, R.H. 2014. *Best Management Practices: Integrated Vegetation Management* (2nd Ed.). Champaign, Illinois: International Society of Arboriculture.

Miller, R.H. 2016. *Pacific power reliability*. Presentation to Operations Staff Meeting. Portland, OR. June 28, 2016.

Miller, R.W., R.J. Hauer, and L.P. Werner. 2015. *Urban Forestry: Planning and Managing Urban Greenspaces* (3rd Ed.). Long Grove, Illinois: Waveland Press, Inc.

Miller, T.L. (Ed.). 1993. *Oregon Pesticide Applicator Manual: A Guide to the Safe Use and Handling of Pesticides*. Corvallis, Oregon: Oregon State University Extension Service.

MindTools. *SWOT Analysis: Discover new opportunities, manage and eliminate threats*. n.d. Retrieved January 9, 2017 from https://www.mindtools.com/pages/article/newTMC_05.htm

Minerals Management Service, Renewable Energy and Alternate Use Program. 2006. *Technology White Paper on Wave Energy Potential on the U.S. Outer Continental Shelf*. Washington, D.C.: United States Department of Interior.

Morris, L.B. 1981. Public relations in utility right-of-way maintenance. *Journal of Arboriculture*, 7:26–28.

Mullen, S. 2013. *Emergency Planning Guide for Utilities* (2nd Ed.). Boca Raton, Florida. CRC Press.

National Fire Protective Association. n.d. *Firestorm '91*. Quincy, Massachusetts: National Fire Protection Association.

National Fire Protection Association. 2006. *Uniform Fire Code*. National Fire Protection Association. Manchester, NH.

National Oceanic and Atmospheric Administration. n.d. (a) *Definition of the NHC track forecast cone*. Retrieved January 12, 2017 from http://www.nhc.noaa.gov/aboutcone.shtml

National Oceanic and Atmospheric Administration. n.d. (b) *What is LiDAR?* Retrieved February 22, 2018 from https://oceanservice.noaa.gov/facts/LIDAR.html

National Oceanic and Atmospheric Administration Earth Observatory. n.d. *Comparing the winds of Sandy and Katrina*. Retrieved January 11, 2017 from http://earthobservatory.nasa.gov/NaturalHazards/view.php?id=79626

National Right to Work Legal Defense and Education Foundation. n.d. *Can I be required to be a union member?* Retrieved February 23, 2017 from http://www.nrtw.org/required-join-pay

National Weather Service. n.d. *Weather Ready Nation: Watch vs. warning.* Retrieved February 21, 2017 from http://www.nws.noaa.gov/os/thunderstorm/ww.shtml

Neal, M. n.d. *System Forester best management practices: Cost drivers* (Utility Arborist Association White Paper). Retrieved March 29, 2018 from https://uaa.wildapricot.org/resources/Documents/Papers/costDriversWP.pdf

New Zealand Parliamentary Counsel Office. 2003. Electricity (Hazards from Trees) Regulations 2003. SR 2003/375. Retrieved January 10, 2017 from http://www.legislation.govt.nz/regulation/public/2003/0375/7.0/DLM233405.html

Next Era Energy Inc. 2013 (April 29). *FPL employees begin company-wide drill to respond to Virtual Hurricane Sheryl.* Retrieved January 12, 2017 from http://nexteraenergy.com/news/contents/2013/042913.shtml

Nichols, D. 1995. *Power Line Fire Prevention Field Guide.* Sacramento: California Department of Forestry and Fire Protection.

North American Electric Reliability Corporation. 2008. *Transmission Vegetation Management NERC Standard FAC-003-2 Technical Reference.* Washington, D.C.: North American Electric Reliability Corporation.

North American Electric Reliability Corporation. 2016. *Transmission Vegetation Management Program Standard FAC-003-4.* Washington, D.C.: North American Electric Reliability Corporation.

Nowak, C.A., and B.D. Ballard. 2005. A framework for applying integrated vegetation management on rights-of-way. *Journal of Arboriculture,* 31:28–37.

Nutter, W. 2012. Safety culture. *Utility Arborist Newsline,* 3:26–28.

Oberbeck, S., and C. Smart. 2004 (January 11). Costly cuts. *Salt Lake Tribune.*

O'Brien, J.G., M.E. Mielke, D. Starkey, and J. Juzwik. 2011. *How to Identify, Prevent, and Control Oak Wilt.* Forest Service Publication NA–FR–01–11. Newtown Square, PA: United States Department of Agriculture.

Occupational Safety and Health Administration (OSHA). n.d. (a) *Elements of an effective safety and health program.* Retrieved January 12, 2017 from https://www.osha.gov/dte/library/safety_health_program/index.html

Occupational Safety and Health Administration (OSHA). n.d. (b) *Guidelines for employers to reduce motor vehicle crashes.* Retrieved January 12, 2017 from https://www.osha.gov/Publications/motor_vehicle_guide.pdf

Occupational Safety and Health Administration (OSHA). n.d. (c) *Hazardous energy control - protective grounding and bonding.* Retrieved February 23, 2018 from https://www.osha.gov/SLTC/etools/electric_power/hazardous_energy_control_protective_ground.html

Occupational Safety and Health Administration (OSHA). 2016. *Worker's rights.* OSHA Publication 3021-11R 2016. Retrieved March 2, 2018 from https://www.osha.gov/Publications/osha3021.pdf

Odom, F.P. 2010, Issue Three. Less is more: Utility line clearance in Tallahassee Florida. *The Council Quarterly,* Retrieved January 12, 2017 from http://www.fufc.org/downloads/councilquarterly10v3.pdf

Oregon OSHA. 2016. *Oregon OSHA's quick guide to safety committees and safety meetings for general industry and construction employers.* Publication 440-0989. Retrieved January 12, 2017 from http://www.cbs.state.or.us/osha/pdf/pubs/0989.pdf

Oregon Secretary of State. 2016. Minimum Vegetation Clearance Requirements. OAR 860-024-0016. Retrieved January 12, 2017 from http://arcweb.sos.state.or.us/pages/rules/oars_800/oar_860/860_tofc.html

Orr, J. 2007. *Impacts of Laws, Regulations and Contracts from a Contractor's Perspective.* Presentation to the Midwestern Chapter of ISA/UAA Utility Vegetation Management Regional Meeting. Kansas City, MO. June 26, 2007.

Ottoboni, A.L. 1997. *The Dose Makes the Poison: A Plain-Language Guide to Toxicity* (2nd Ed.). New York: John Wiley & Sons.

Outlaw, W. 1998. *Smart Staffing: How to Hire, Reward, and Keep Top Employees for Your Growing Company.* Chicago: Upstart Publishing Company.

Pacific Gas and Electric Company. n.d. *Right tree right place.* Retrieved January 11, 2017 from https://www.pge.com/en_US/safety/yard-safety/powerlines-and-trees/right-tree-right-place/right-tree-right-place.page

Pacific Northwest Center of Excellence for Clean Energy / "A Centralia College Partnership." 2012. *Skill standards for utility customer service representatives.* Washington State University Extension Energy Program.

Patel, M.R. 2006. *Wind and Solar Power Systems: Design, Analysis, and Operation.* New York: Taylor & Francis Group.

Perry, P.B. 1977. Management's view of the tree trimming budget. *Journal of Arboriculture,* 3:157–160.

Petersen, D. 1997 (January). Behavior-based safety systems: A definition and criteria to assess. *Professional Safety,* January 1997.

Petersen, D. 2001. *Safety Management: A Human Approach* (3rd Ed.). Des Plaines, Illinois: American Society of Safety Engineers.

Porter, P., and N. Cohn. 2017. *CNUC Transmission Pipeline Vegetation Management Survey 2017.* West Des Moines, IA: CNUC.

Professional Training Systems, Inc. 1979. *Electric Utility System Operation.* Portland, Oregon: Professional Training Systems, Inc.

Public Relations Society of America. 2005. Restoring a utility's wounded reputation. Retrieved January 13, 2017 from http://www.prsa.org/searchresults/view/6bw-0502b08/0/restoring_a_utility_s_wounded_reputation?utm_source=prsa_website&utm_medium=case_studies&utm_campaign=business_case#.WHj7t00zXSc

Public Relations Society of America. 2012. *Public Relations Defined: A Modern Definition for the New Era of Public Relations.* Retrieved March 1, 2018 from http://prdefinition.prsa.org/

Public Utility Commission of Oregon. 2016. *Laws Relating to the Public Utility Commission of Oregon.* Salem, Oregon: Public Utility Commission. Retrieved January 12, 2017 from http://www.puc.state.or.us/commission/docs/lawbook2016.pdf

Puget Sound Energy. 2010. *Vegetation management PSE: Keeping trees clear for safety and reliability* (PSE Fact Sheet). Retrieved January 11, 2017 from http://pse.com/aboutpse/PseNewsroom/MediaKit/017_Vegetation_Management.pdf

Rachlin, R. 1999. *Handbook of Budgeting* (4th Ed.). New York: John Wiley and Sons, Inc.

Ramirez, M. n.d. Better online communication makes business sense. Retrieved January 9, 2017 from http://www.businessknowhow.com/internet/communication.htm

Reason, J. 2000. Human error: Models and management. *Western Journal of Medicine,* 172:393–396.

Redding K.D., P.L. Burch, and K.C. Miller. 1994. Growth, biomass, and trim/chip time reduction following application of flurprimidol tree growth regulator. *Journal of Arboriculture*, 20:38–45.

Rees, W.T., T.C. Birx, D.L. Neal, C.J. Summerson, F.L. Tiburzi, and J.A. Thurber. 1994. *Priority trimming to improve reliability*. Retrieved January 10, 2017 from http://www.ecosync.com/tdworld/PRIOTRIM03.pdf

RightNow Technologies. 2010. *Customer experience report North America 2010*. Retrieved March 7, 2018 from http://www.overells.com.au/files/docs/customer_experience_ impact_north_america.pdf

Rodriguez, J. 2017. *The pros and cons of the Right-to-Work Act*. Retrieved April 5, 2018 from https://www.thebalance.com/right-to-work-act-844515

Rollins, K.E., D.K. Meyerholz, G.D. Johnson, A.P. Capparella, and S.S. Lowe. 2012. A forensic investigation into the etiology of bat mortality at a wind farm: Barotrauma or traumatic injury? *Veterinary Pathology*, 49:632–371.

Ross, T. 2011. From suburban settings to mountainous terrain, selecting the right technology to help strengthen efficiency and increase safety in any environment. *Utility Arborist Newsline*, 2:16–17.

Russell, D.B. 2011. *Best Practices in Vegetation Management for Enhancing Electric Service in Texas. Public Utility Commission of Texas Project 38257*. College Station: Texas Engineering Experiment Station.

Rymer, R. 2008. Reuniting a river: Unlikely allies work to let the Klamath river run free. *National Geographic*, 214:134-155.

Salmone, J.P., and P.T. Pons (Eds.). 2007. *PHTLS Prehospital Trauma Life Support* (6th Ed.). St. Louis, Missouri: Mosby Elsevier.

Sanchez. J. 2012. Electrocution. *Utility Arborist Newsline*, 3:24.

Sankowich, S. 2013, July/August. The vegetation management program. *Electricity Today*, 26(6).

Satterfield, D. 2013. Hurricane forecasts and the shrinking cone of uncertainty. [American Geophysical Union Blogosphere]. Retrieved January 10, 2017 from http://blogs.agu.org/wildwildscience/2013/04/05/hurricane-forecasts-and-the-shrinking-cone-of-uncertainty/

Schewe, P.E. 2007. *The Grid: A Journey Through the Heart of our Electrified World*. Washington, D.C.: Joseph Henry Press.

Schwartz, R.M. 2006. *The Legal Rights of Union Stewards* (4th Ed.). Cambridge, Massachusetts: Working Rights Press.

Scientific American. 1999. *Science Desk Reference*. New York: John Wiley and Sons.

Scott. J. 2012. How's my driving? *Utility Arborist Newsline*, 3:18–24.

Senseman, S.A. (Ed.). 2007. *Herbicide Handbook* (9th Ed.). Lawrence, Kansas: Weed Science Society of America.

Shigo, A.L. 1986. *A New Tree Biology and Dictionary: Facts, Photos and Philosophies on Trees and Their Problems and Proper Care*. Durham, New Hampshire: Shigo and Trees, Associates.

Shigo, A.L. 1990. *Pruning Trees Near Electric Utility Lines: A Field Pocket Guide for Qualified Line-Clearance Tree Workers*. Durham, New Hampshire: Shigo and Trees, Associates.

Shigo, A.L. 1991. *Modern Arboriculture: A Systems Approach to the Care of Trees and Their Associates*. Durham, New Hampshire: Shigo and Trees, Associates.

Shim, J.K., and J.G. Siegel. 1994. *Budgeting Basics and Beyond: A Complete Step-By-Step Guide for Nonfinancial Managers*. Paramus, New Jersey: Prentice Hall.

Simpson, P., and R. Van Bossuyt. 1996. Tree-caused electric outages. *Journal of Arboriculture*, 22:117–121.

Smiley, E.T., N. Matheny, and S. Lilly. 2017. *Best Management Practices: Tree Risk Assessment*. (2nd Ed.). Champaign, Illinois: International Society of Arboriculture.

Smith, E., R.J. Arsenault, and C. Branch. 2014 (April). The long road to improved storm planning begins here. *FTI Journal*. Retrieved January 11, 2017 from http://www.ftijournal.com/uploads/pdf/fti_energy_revised.pdf

Smith, R.L. 1980. *Ecology and Field Biology* (3rd Ed.). New York: Harper & Row.

Society for Human Resource Management. 2013. *Making safety committees work.* Retrieved January 11, 2017 from http://www.shrm.org/hrdisciplines/safetysecurity/articles/Pages/Workplace-Safety-Committees.aspx

Spalding, J.R. n.d. *Grievance mediation*. Retrieved March 20, 2018 from http://www.michigan.gov/documents/lara/GRIEVANC_mediation_article_Mar_2017_556505_7.pdf

Spitzer, D.R. 1995. *SuperMotivation: A Blueprint for Energizing Your Organization From Top to Bottom*. New York: American Management Association.

State Compensation Insurance Fund. n.d. *Employee safety responsibilities*. Retrieved January 11, 2017 from http://content.statefundca.com/safety/safetymeeting/SafetyMeetingArticle.aspx?ArticleID=232

Stedman, J., and R. Brockbank. 2012. Integrated vegetation management on pipeline rights-of-way: Part one. *Utility Arborist Newsline,* 3:1,4–5.

Sullivan, M., and J. Schellenberg. 2011, April. Smart grid economics: The cost-benefit analysis. *Renew Grid,* 2(3).

Swann, P. 2014. *Cases in Public Relations Management: The Rise of Social Media and Activism* (2nd Ed.). New York: Routledge.

Tate, R.L. 1986. Contracting for city tree maintenance needs. *Journal of Arboriculture*, 12:97–100.

Tennessee Valley Authority. 2011. *Human performance program*. Chief Operating Officer Standard Programs and Processes. COO-SPP-03.1.14.

Tennessee Valley Authority. 2017. *Raccoon Mountain*. Retrieved December 12, 2017 from https://www.tva.gov/Energy/Our-Power-System/Hydroelectric/Raccoon-Mountain

Texas State Historical Association. n.d. *Texas Almanac: City population history from 1850–2000*. Retrieved February 21, 2017 from http://texasalmanac.com/sites/default/files/images/CityPopHist%20web.pdf

Tomasovic, B.S. 2011. A high-voltage conflict on Blackacre: Reorienting utility easement rights for electric reliability. *Columbia Journal of Environmental Law,* 36:1–57.

Tomosho, R. 1998, November 17. Alex Shigo lops off big branches, and his critics call him crazy. *The Wall Street Journal*. Retrieved January 9, 2017 from http://www.wsj.com/articles/SB911256283673984500

Tracy, J.A. 2008. *Accounting for Dummies* (4th Ed.). Hoboken, New Jersey: Wiley Publishing.

United States Department of Energy, Office of Electricity Delivery and Energy Reliability. 2012. *A review of power outages and restoration following the June 2012 derecho*. Retrieved January 9, 2017 from http://www.oe.netl.doe.gov/docs/Derecho%202012_%20Review_080612.pdf

United States Department of Transportation, Federal Highway Administration, Office of Operations. n.d. *Glossary: simplified guide to the Incident Command System for transportation professionals*. Retrieved January 9, 2017 from http://ops.fhwa.dot.gov/publications/ics_guide/glossary.htm

United States Department of Transportation, Federal Highway Administration. 2008. *Vegetation control for safety: A guide for local highway and street maintenance personnel.* Publication No. FHWA-SA-07-018. Washington, D.C.: U.S. Department of Transportation.

United States General Services Administration. 2005. *Guide to Federal Fleet Management.* Washington, D.C.: United States General Services Administration.

Utility Arborist Association. 2009. *Utility best management practices tree risk assessment and abatement for fire-prone states and provinces in the western region of North America.* Champaign, Illinois: Utility Arborist Association. Retrieved January 11, 2017 from http://www.novembriconsulting.com/images/TreeRiskAssessmentBMP_20090629.pdf

Utility Arborist Association. 2014. *Scheduling, optimum cycle, and the cost of deferring maintenance.* UAA White Paper.

UN Atlas of the Oceans. *Human settlements on the coast.* Retrieved March 1, 2018 from http://www.oceansatlas.org/subtopic/en/c/114/

University of Illinois Atmospheric Sciences. n.d. *Freezing rain.* Retrieved January 9, 2017 from http://ww2010.atmos.uiuc.edu/(Gh)/guides/mtr/cld/prcp/zr/frz.rxml

US Legal, Inc. n.d. *Collective bargaining law and legal definition.* Retrieved January 9, 2017 from https://www.uslegal.com/c/collective-bargaining/

Van Soelen, W. 2006. *The Guidebook for Linemen and Cablemen.* Clifton Park, New York: Delmar Cengage Learning.

Vera-Art, S. 2017 (Jan/Feb). An organic ROW restoration project. *Utility Arborist Newsline*, Volume 8, No. 1, pp. 1–3,5.

Walsh, B. 2013, August 12. 10 Years after the great blackout, the grid is stronger — but vulnerable to extreme weather. Time.com. Retrieved January 6, 2017 from http://science.time.com/2013/08/13/

Walsh, T. 2012. Falls. *Utility Arborist Newsline,* 3:1–4.

Webster, J., L. DiRuzza, and M. Deric. 2011 (January). Vision vs. reliability: Finding the balance. *Transmission & Distribution World,* 63.

Weik, K.E., and K.M. Sutcliffe. 2007. *Managing the Unexpected.* Hoboken, New Jersey: John Wiley & Sons, Inc.

Wellman, W.R. 1959. *Elementary Electricity.* New York: D. Van Nostrand Company.

Wenderlich, R.L. 1997. *The ABCs of Successful Leadership: Proven, Practical Attitudes, Behaviors and Concepts Based on Core Values That Result in Successful Leadership.* Ellicott City, Maryland: Success Builders Incorporated.

Whitaker, J.C. 1999. *AC Power Systems Handbook* (2nd Ed.). New York: CRC Press.

World Meteorological Organization. n.d. *Tropical cyclone naming.* Retrieved January 9, 2017 from http://www.wmo.int/pages/prog/www/tcp/Storm-naming.html

Xcel Energy Inc. 2007. *Understanding easements and rights-of-way.* Retrieved January 9, 2017 from https://www.xcelenergy.com/staticfiles/xe/Corporate/ Corporate%20PDFs/SLK_ROWBrochure_FS.pdf

Yahner, R.H., and R.J. Hutnik. 2004. Integrated vegetation management on an electric transmission right-of-way in Pennsylvania. U.S. *Journal of Arboriculture,* 30:295–300.

Yang, M.J., D.L Zhang, and H.L. Huang. 2008. A modeling study of Typhoon Nari (2001) at landfall. Part 1: Topographic effects. *Journal of the Atmospheric Sciences,* 65:3095–3115.

Ziesemer, D. 1977. Determining needs for street tree inventories. *Journal of Arboriculture,* 4:208–213.

OTHER RESOURCES

National Right to Work Legal Defense and Education Foundation, Inc.
http://www.nrtw.org

Reliabilitree. Westar Energy
https://www.westarenergy.com/reliabilitree-program

Share the Space: Line Clearing and Safe Reliable Electric Service. Consumers Energy
http://www.consumersenergy.com/uploadedFiles/CEWEB/SHARED/Forestry-Brochure.pdf

Trees and Rights of Way. Duke Energy
https://www.duke-energy.com/community/trees-and-rights-of-way/what-can-you-do-in-right-of-way/how-we-manage-rights-of-way

Vegetation Management. Seattle City Light
http://www.seattle.gov/light/vegmgmt/

Weather Ready Nation. National Oceanic and Atmospheric Administration, National Weather Service
http://www.nws.noaa.gov/com/weatherreadynation/

ANSWERS TO WORKBOOK QUESTIONS

Chapter 1 Safety

Fill in the Blank
1. Occupational Safety and Health Administration (OSHA)
2. act, condition
3. two-minute
4. 10, 3.05
5. 1,000, 1,832
6. direct contact, indirect contact
7. touch potential
8. step potential
9. indirect contact

Multiple Choice
1. a
2. a
3. a
4. c
5. d
6. c

Chapter 2 Program and Personnel Management

Fill in the Blank
1. written, verbal
2. strategic, tactical
3. specifications
4. three, single
5. just-in-time
6. work breakdown
7. cost centers
8. capital, operating
9. low, high
10. unit price
11. straight-line, accelerated

Multiple Choice
1. c
2. a
3. d
4. b
5. d

Chapter 3 Utility Pruning

Fill in the Blank
1. easement, right-of-way
2. maintenance interval
3. tree owner
4. response growth
5. greater
6. subordination
7. epicormic
8. increase
9. preventive, reactive

Multiple Choice
1. d
2. a
3. c
4. d
5. d
6. b
7. b

Copyright © 2018 by International Society of Arboriculture. All rights reserved.

Chapter 4 Integrated Vegetation Management

Fill in the Blank

1. wire zone, border zone
2. poison
3. incompatible plants
4. specific, measurable, achievable, relevant, timely
5. imminent, probable
6. action thresholds
7. mechanical
8. plant and wildlife, soil
9. acute, chronic
10. 100, 30
11. chemical side-pruning
12. biological control
13. wire zone

Multiple Choice

1. d
2. d
3. d
4. a
5. b
6. a
7. c
8. c
9. d
10. b
11. d
12. b

Chapter 5 Electrical Knowledge

Fill in the Blank

1. amp
2. 120, 765
3. fault
4. substations
5. voltage
6. insulators, conductors
7. tree wire
8. circuit breakers
9. back feed
10. voltage gradient, stem diameter, species
11. frequency, duration
12. mechanical tear down
13. interactivity, blackouts
14. minimum vegetation clearance distance

Multiple Choice

1. c
2. b
3. c
4. d
5. a
6. b

Chapter 6 Storm Preparation and Response

Fill in the Blank
1. sleet, freezing rain
2. incident command system
3. emergency operations center
4. home pay, local pay
5. 67
6. brush
7. stand-down
8. emergency

Multiple Choice
1. d
2. a
3. a
4. c
5. b
6. d
7. a

Chapter 7 Communications

Fill in the Blank
1. public safety
2. stakeholders
3. communications
4. proactive
5. active listening
6. desensitization
7. jargon
8. underestimate
9. press releases

Multiple Choice
1. c
2. b
3. c
4. c
5. c

INDEX

Page numbers in *italics* indicate tables. Page numbers in **bold** indicate figures.

A

AC. *See* alternating current
accelerated depreciation, 37–38
accident pyramid, 8–**9**, 217
accidents, 4, 217
accrual accounting, 39, 217
action thresholds, 82, 94–95, 217
active listening, 8, 209–**210**, 217
acute toxicity, *110*, 111, 217
aerial application, *102*, 104, 217
aerial lifts, 4, 96, **97**, 179, 217
AHAS inhibitors, 113, *114*, 217
airbrake switches, 143, **144**, 217
allelopathy, 104, 217
ALS inhibitors, 113, *114*, 217
alternating current (AC), 126, 129, 133–**134**, 217
aluminum conductor steel-reinforced cable (ACSR), 140, 217–218
ampere (amp, A), 126, **127**, 218
angry customers, 211–212
ANSI A300
    Part 1, mechanical pruning, 99
    Part 7, integrated vegetation management, **83**, 96, 155, 218
    Part 9, tree risk assessment, 92, **93**, *93*
ANSI Z133
    definition, 218
    electrical safety standards, 14–16, *15*
    employer responsibilities, 5, 15
    incidental vs. qualified line-clearance arborists, *15*–16
approach distances, *15*–*16*, *17*, 20, 218, 227
archeological sites, 91
automatic line reclosers, 145–**146**, 218
auxin transport inhibitors, 218
auxins, 218
    synthetic auxins, 110, 113, *114*, 233

B

back feed
    definition, 20, 150, 218
    safety precautions, 20–21
    storm response, 186
bare-ground treatments, **104**
bargaining unit, 50, 218
basal applications, *102*–*103*, 218
basic assessment, 92–94, *93*, 226
behavior-based incident prevention
    definition, 218
    high-reliability organizations, 12–13
    multiple causation theory, 10–11, *12*
    punishment for incidents, 10, 21
    safety, 8–13, **9**
    safety culture, 5, 13, **21**–22, 230
best practices, 60
    Best Management Practices: Integrated Vegetation Management, 156
    Best Management Practices: Tree Pruning, 61
    Best Management Practices: Tree Risk Assessment, 94
    Best Management Practices: Utility Pruning of Trees, 96
    integrated vegetation management, 82
    Utility Best Management Practices Tree Risk Assessment and Abatement, 94
biological control, 104–105, 218
Blair, G.D., vii, ix
border zone, **85**–**86**, 218
    pipe zone-border zone, **88**, 229
bracket grounding, 20
branch bark ridge, **63**, 218
branch collar, **63**, 218
branch removal cut, **63**, 219
branch unions, 66, 219
brand (business), 199, **204**, 219
broadcast applications, *102*, 103–104
brush cutters, 96, 219
budgets, 36–39, 219, 229
    justification, 199–200
bus work, 219
business continuity, 173, 219

C

cable television (CATV) lines, 141, **143**
call-out lists, 173
"Can't do–Won't do" test, 5
capacitive reactance, 127, 219
capacitors, **151**, 219
capital budgets, 37–38, 219
cascading outages, 154, **155**, 219
chain of command, 172–173, 219
checklists, 5, 16, **18**
chemical control
    chemical pruning, 77, 219
    definition, 219
    herbicide, 112–**113**, 224
    herbicide application methods, 101–**104**, **102**, *102*, **103**
    herbicide closed chain of custody, 113–**118**, **115**, **116**, **119**, 219
    herbicide mode of action, 113, *114*, 227
    herbicide recordkeeping, 114, 117, **118**
    inventory management, 117–**118**
    selectivity, 110
    tree growth regulators, 72, 99–**101**, *100*, 112, 234
chemical pruning, 77, 219
chemicals. *See also* chemical control
    half-life, 111–112
    labels, 109–*110*, 111, **115**, **118**, 225
    selectivity, 110
    toxicity, *110*–112, 233
chronic toxicity, 110, 219
circuit, 127, 219
circuit breakers, 144–145, **146**, 219
circuit protection, 143–148, **146**, **147**
circuit work, 32, 219
clearances
    deferring maintenance, 72, **74**, 221
    definition, 219
    following work, 108
    minimum vegetation clearance distance, 95, 155, 227
    regulations, 154–*157*
    rural vs. urban areas, **30**
    specification expectations, 43
    Transmission Vegetation Management Program, 83
    utility pruning objectives, 67–**69**

close calls
    300 per major incident, 9, 11
    as leading indicators, **9**
    reporting, 8, 10, 12, 21
closed chain of custody, 113–**118**, **115**, **116**, 219
codominant branches/stems, **63**, 219
combined-cycle system, 135, 219
common interest, 220
communication
    active listening, 8, 209–**210**, 217
    angry customers, 211–212
    chain of command, 172–173
    close call reporting, 8, 10, 12, 21
    crew accommodation services, **182**
    crew safety, 8
    crew training in, 189, 197, 199, 200, 202, 205
    customer relations, **197**–199, **198**
    customer vs. public relations, **204**–205
    customer-friendly language, 211, 223
    emergency operations center, **173**
    Incident Command System, 165, 172–173
    motivation, *49*, 200
    outreach methods, 205–**207**, **206**, 228
    phone etiquette, 210, 211
    post-job review, 8, 190
    pre-storm check, 173
    quality assurance, 109
    safety committee, 6
    specifications, 28–29, 42–43, 232
    stakeholder interests, 199–205, *201*, **202**, **203**
    storm brush disposal, 183–**184**
    storm response crews, 186, 189
    storm response debriefing, 190
    storm response interruptions, 184
    storm response media relations, 189
    support for utility pruning, 76, 196–197, **198**, 199, *201*, **210**
    talking with customers, **207**–212, *208*, **210**
    targeted messaging, 199–205, *201*, **202**, **203**, 227, 233
    tree beyond scope of assignment, 59, 60
community meetings, **207**
compact construction, **105**, *107*
compartmentalization of decay in trees (CODIT), 73
compatible vegetation, 82, 84, 220
comprehensive evaluations, 89–**90**, 220
conductivity, 127, 220

conductor sag, 140, **155**, 220
conductors, 140–**143**, **142**, 220, 222
constraint triangle, **32**, 220
contractors
    budgetary reports, 39
    crew communication training, 189, 197, 199, 200, 202, 205
    definition, 220
    fleet management, 45–**47**
    stakeholder interests, 199–200
    storm preparation, 177
    storm response, 180, 183, 188
    talking with customers, **207**–212, *208*, **210**
    use of, 40
contracts
    contract structures, 40–*44*, **41**, *42*
    definition, 40, 220
    mediation in disputes, 51
    procurement, 44–45
    requests for proposals, 45
    specifications, 28–29, 232
    specifications and lump sum, 43, *44*
    specifications and unit price, 42–43
    stakeholder satisfaction, 200
control methods
    biological, 104–105, 218
    chemical. *See* chemical control
    debris disposal, 108–109
    definition, 220
    engineering solutions, **105**–*108*, **106**, *107*
    implementation, 108–109
    manual, **95**–96, **97**, 226
    mechanical, **76**, 96–99, **97**, **98**, 226
    quality assurance, 109
coppicing, 220
Coriolis force, 166–**167**, 220
corrosiveness, 110, 220
cost centers, 36, 220
cost types, 36
covered overhead primary wire, **105**, *107*, 140, 234
cover-type conversion, 104, 220
cover-type mapping, **89**, 220
crews, 179–180, 183–185, **184**, 220
    accommodations, 180–**181**, **182**, 184–185, 186
    communication training, 189, 197, 199, 200, 202, 205
    talking with customers, **207**–212, *208*, **210**
crisis management, 31, 221
critical path, 33, 221
critical service provider, 165, 221
cultural control methods, 105, 221
cultural site protections, 91
current, 126, **127**, 221
current surges, 128, 221
customer relations
    angry customers, 211–212
    communication importance, **197**–199, **198**
    customer dissatisfaction, **198**–199
    customer perception, **198**
    customer-friendly language, 211, 223
    definition, 221
    outreach methods, 205–**207**, **206**, 228
    public relations versus, **204**
    storm brush disposal, 183–**184**
    support for utility pruning, 76, 196–197, **198**, 199, *201*, **210**
    talking with customers, **207**–212, *208*, **210**
    targeted messaging, 200–205, *201*, **202**, **203**
    work on display, **197**
customer service, **204**–205, 221
    talking with customers, **207**–212, *208*, **210**
cutouts, **147**, 221
cycles
    action thresholds and, 94
    definition, 221
    maintenance intervals, 29–30, 226
cyclones, 166–*168*, *167*, **174**–175, 221, 224, 234
    tornados versus, **167**, 169
    tracking, **178**–**179**

D

data. *See* documentation
debris disposal, 108–109
    storm response, 183–**184**, 186
decision packages, 38, 221
defects, 67, 92, 165, 166, 221
deferring maintenance, *72*, **74**, 221
delta construction, 141, 221
dependencies, 33, 221
depreciation, 37–38
derecho, **169**, 221
desensitization, 210, 221
direct contact, 16, 19, 20, 221

direct costs, 36
directional pruning, **70–71**, 221
    contributions of Alex Shigo, **73**
disaster, 185–186, 221. *See also* storm response
disaster declaration, 186, 221
disaster management cycle, **164**
discipline for incidents, 10, 21
disconnect switches, 143, **144**, 221
discretionary dependencies, 33
distribution network, **131**–133, **132**, 222
    circuit protection, 145–148, **146**, **147**
    distribution lines, **131**, 141, 222
    distribution systems, 132–133
    electrical system layout, **129**
    feeders, 127
    guy wires, 141, **142**, 224
    other equipment, *148*–*151*, **149**, **150**
    poles, 139, **140**, 141
    switches, 143, **144**
    transformers, **132**, 141, 148, **150**–151, 233
    underground, 105–106, *107*, **143**
documentation
    herbicide recordkeeping, 114, 117, **118**
    IVM strategies, 83
    labor grievances, 51
    SAIDI, 151–152, 231
    SAIFI, 151–152, 231
    storm response, 188
    tree risk assessment, 92
door hangers, 205, **206**
dose, *110*, 222
down guy, *107*, 141, **142**, 222
downed lines, 20

E

early successional plant community, 84, 104, 113, 222
earthing (grounding), 20, 224
easements, 59–**60**, 222
electric grid, **129**, 222
    circuit protection, 143–**148**, **146**, **147**
    conductors, 140–**143**, **142**, 220, 222
    distribution. *See* distribution network
    generation. *See* generation of electricity
    transmission. *See* transmission of electricity
    tree-caused interruptions, 151–**154**, **153**

electric shock
    definition, *14*, 222
    tree on line, 19
    workplace trauma, 4, 13, *14*
electric utility, 222. *See also* utility facilities
    utilities as stakeholders, 199, 200
electrical conductor, 140–**143**, **142**, 220, 222
electrical potential (voltage), 126, 222
electrical safety standards, 14–**16**, *15*, 156–157, 227
"electrical trimming," 153, **154**
electricity fundamentals, 126–128, **127**
electrocution, *14*, 222
electromagnetic field, 128, 222
electromotive force, 126, 222
email for outreach, 206
emergency declarations, 185–186
emergency operations center (EOC), **173**, 177
    crews, 177, 179–180, 220
emotional intelligence, 48–49
empathy, 212, 222
employee responsibilities
    communication training, 189, 197, 199, 200, 202, 205
    performance evaluation, 49–50
    professional development, 200
    safety, 7, 8, 11, **21**
employees as stakeholders, 199–200
employer responsibilities. *See* personnel management
energized conductor, 140, 222
    storm response, 186
    telephone and cable TV lines, **143**
    treat all wires as, 141, 186
engineering for control, **105**–108, **106**, *107*
Enhanced Fujita Scale, *170*, 222
entrepreneurial budgeting, 38
environmental protection
    control methods and, **95**, 96
    explanation of, 91
    mixing herbicide, 117
    stakeholder targeted messaging, *201*–**202**, **203**
epicormic shoots, **66**, 70, 222
EPSP inhibitors, 113, *114*, 222
evergreen contracts, 40, 222
excurrent growth habit, **65**, 70–71, 222
exposure, 4, *110*, 111, 222
    closed chain of custody, 114, **116**–117

external dependency, 33
external stakeholder messaging, 200–**203**, *201*, *202*, 222
extratropical cyclones, 168, 223

F

FAC-003 (NERC; 2008), 94, 95, 154–155, 223
failure modes, 152–**153**, 223
falls (workers), 4, 13–14
Faraday, James, 128, 148
fate in soil, 111–112, 223
faults, 128, 223
feeder, 127, 145, 223
feller-buncher, 96, **98**
field windings, **133**, 134, 223
finish-to-start dependency, 33
fire rules and regulations, 157–158
first impression, 207–*208*
first responders, 164, **165**, 223
fixed assets, 37
fixed costs, 36
fixed mindsets, 48
fixed-price contracts. *See* lump-sum contracts
fleet management, 45–**47**, 223
flush cuts, **63**
forecasting revenue, 38
fossil fuel steam turbine plants, **133**, 134–135
freezing rain, 170, **171**, 223
Frequently Asked Questions (FAQ) pages, 206
friendly language, 211, 223
frill treatments, *102*, 223
fronds, **74**, 75, 223
funnel clouds, 169, 223
fuses, **147**, 223

G

Gantt charts, 33–**34**
gas turbines, 135, 223
generation of electricity
    electrical system layout, 128–**129**, *129*
    gas turbines, 135, 223
    generators, **133–134**
    geothermal, 138
    hydroelectric, 135, **136**, 224
    ocean current, 139
    solar, **137–138**, 232
steam turbines, **133**, 134–135
wind turbines, 135–137, **136**
geographic information system (GIS), **35**–36, 223
    cover-type mapping, **89**
    herbicide application, 117
    quality assurance, 109
geothermal energy, 138
GIS. *See* geographic information system
global positioning system (GPS)
    crews in storm response, 183, **184**
    definition, 223
    vehicle monitoring systems, **47**
    work planner capabilities, 34
goal setting, 49–50
    SMART goals, 6, 50, *84*, 231
GPS. *See* global positioning system
Graham cover-type mapping, **89**, 223
gravitational energy, 13
greenspace inventories, 88–89
grid work, 32, 223
grievances, 51
ground fault, 16, 140
ground wires, 140, 223
grounded, 21, 141, 223
grounding, 20, 224
ground-to-conductor clearance, 224
growth mindsets, 48
guy wires, 141, **142**, 224

H

hack and squirt, *102*, 223
half-life, 111–112, 224
hazard tree, 42, 224
heading cut, **63**, 64, 224
heavy, wet snows, 171, **172**, 224
Heinrich, Herbert, 8–**9**, 217, 218
helmets, 13
Hendrix spacer system, 105, **106**, *107*
herbicides, 112–113, 224
    application methods, 101–**104**, *102*, 102, **103**
    application via GIS, 117
    closed chain of custody, 113–**118**, **115**, **116**, 219
    inventory management, 117–**118**
    mode of action, 113, *114*, 227
    recordkeeping, 114, 117, **118**
    selectivity, 110

Hertz (Hz), **134**, 224
high winds, 172
high-reliability organizations, 12–13, 224
high-voltage lines, 126, *129*, 224
hotspotting, 31, 224
human error, 4, 8
hurricanes, 166–*168*, **167**, **174**–175, 221, 224, 234
    tracking, **178–179**
hydration (crews), 185, 186, 224
hydroelectric generation plants, 135, **136**, 224

I

ice-laden branches, 170, **171**
ICS. *See* Incident Command System
imminent threat, 94, 224
impedance, 127, 224
Incident Command System (ICS), 165, 172–173, 224
incidental line-clearance arborists, *15–16*
incidents
    axioms of safety, 11
    definition, 4–5, 224
    multiple causation theory, 10–11
    punishment for, 10, 21
    trauma, 13–*14*
    utility pruning risk management, 58
incompatible plant species, 82, 84, 224
    tolerance levels, 82, 94–95, 233
indirect contact, 16, 19, 20, 224
indirect costs, 36
inductance, 128, 148, 225
inductive reactance, 127, 225
initial clearing, 110, 225
insulators, *148*, **149**, 225
    conductivity, 127
    phases mounted on, **132**, 141
integrated pest management, 82, 225
integrated vegetation management (IVM)
    action thresholds, 94–95
    control method implementation, 108–109
    control method selection, **95**–*108*, **97–106**, *107*
    debris disposal, 108–109
    definition, 82, 225
    environmental protection, 91
    flowchart of process, **83**
    generating support for, 76, 196–197, **198**, 199, *201*, **210**
    mitigation plan, 92, 94
    objectives, **83**, *84–***90**, **85–89**, 95
    quality assurance, 109
    regulations, 154–158
    scale of plan, 88–**90**, **89**
    site evaluation, 90–94, **93**, *93*
interconnection reliability operation limit (IROL), 130, 154–155, 225
intermittent faults, 128, 225
internal stakeholder communication, 199–200, 225
interruptions, 126, 225
    costs of storm-caused, 164
    tree-caused, 151–**154**, **153**
IROL. *See* interconnection reliability operation limit
irreversibilitiy, 111, 225
IVM. *See* integrated vegetation management

J

job briefing, 8, 16, **17**, 225
    checklist for, **18**
    crews in storm response, 183, 186
just-in-time management, 31, 225

K

key performance indicator (KPI), 50, 225
kilovolt, 126, 225
kinetic energy, 13

L

labels on chemicals, 109–*110*, 225
    returnable, reusable containers, **115**, **118**
    toxicity information, *110*, 111
labor unions, 50–52, 173, 226
lagging indicators, **9**–10, 226
landowner targeted messaging, *201*–**202**, **203**
large-scale responses, 180–185, **181**, **182**, **184**, 226
lateral, 64, 70, 145, 226
lateral bud, 64, 226
LC$_{50}$, *110*, 111, 226
LD$_{50}$, *110*, 111, 226
leadership skills, 48–50, *49*
leading indicators, **9**–10, 226
level one assessment, 92–94, **93**, 226

level two assessment, 92–94, *93*, 226
lever arm, 166, 226
LiDAR (light detection and ranging), 89–**90**, 226
lightning arrestors, 147–**148**, 226
limited visual assessment, 92–94, **93**, 226
line sag, 140, **155**, 220
line sectionalizers, 146, **147**, 226
line-item budgeting, 38
lines
    definition, 226
    downed, 20
    engineering solutions, **105**–*108*, **106**, *107*
    protective grounds, 20
    tree on line, 19, **187**
    underground, 105–106, *107*, **143**
lion tailing, 67, 226
loads on trees, 165–**166**, 226
    winter storms, 170–**171**, *172*
lockout, 145, 226
loop primary circuit, 133
loop radial distribution systems, 132
low-voltage lines, *129*, 226
lump-sum contracts, 40, **41**, *42*, 43–44, 226

M

maintenance budgets, 38
maintenance of fleet, 46–**47**
maintenance/management intervals, 29–30, 226
    action thresholds and, 94
    deferring maintenance, *72*, **74**, 221
    utility pruning intervals, 71–**74**, *72*
maliciousness, 4
management. *See* program management
mandatory dependency, 33
manual control, **95**–96, **97**, 226
mechanical control, **76**, 96–99, **97**, **98**, 226
mechanical failure interruption, 152
mechanical pruning, **76**, 96–99, **97**, **98**, 226
media relations
    customer dissatisfaction, 199, 202
    definition, 227
    as external stakeholders, 200–**203**, *201*, **202**
    outreach methods, 205–**207**, **206**, 228
    press releases, 207, 229
    storm response, 189
    targeted messaging, *201*, **202–203**

mediation in disputes, 51
megawatt, 127, 227
meristem, 153, 227
messaging, 199–205, *201*, **202**, **203**, 227, 233
mid-cycle pruning, 30, 227
mindset, 48
minimum approach distances, *15–16*, *17*, 20, 227
minimum vegetation clearance distance (MVCD), 95, 155, 227
mitigation plan, 92, 94
mobilization, 227. *See also* storm preparation; storm response
mode of action, 113, *114*, 227
momentary interruptions, 145–146, 227
monocotyledons (monocots), **74**, 227
motivation, *49*, 200
multiple causation theory, 10–11, 227
mutation, 111, 227
mutual assistance program, 177, 227

N

National Electrical Safety Code (NESC), 156, 227
National Hurricane Center (NHC), **178–179**
National Labor Relations Act (1935), 50, 227
National Land Cover Database, **89**
natural pruning, 61, 227
    contributions of Alex Shigo, **73**
    directional pruning, **70–71**, 221
NERC. *See* North American Electric Reliability Corporation
neutral wires, 140, 227
New Zealand regulations, *157*
news media, 227. *See also* media relations
node, 64, 228
non-damaging level, 94, 228
non-selective herbicides, 110, 228
North American Electric Reliability Corporation (NERC), 83, 154, 228. *See also* FAC-003
nuclear steam turbine plants, **133**, 134–135

O

Occupational Safety and Health Act (OSHA; 1970), 5, 6–7, *15–16*, 228
Occupational Safety and Health Administration (OSHA), 5, 228

ocean current generation, 139
ohm (Ω), **127**, 228
Ohm's law, **127**
oil switches, 143, 228
oncogenicity, 111, 228
on-condition maintenance. *See* predictive maintenance
operating budgets, 38, 228
operation, 144, 228
outages, 128, 228
    costs of storm-caused, 164
    distribution systems, 132–133, 145
    system reliability indices, 151–152
    tree-caused, 151–**154**, **153**
outreach, 205–**207**, **206**, 228
overcurrent, 128, 143–144, 228
overhanging branches, 71, **72**

P

palm pruning, **74**–75
partial tree evaluations, 94
performance appraisal, 49–50, 228
performance budgeting, 38, 228
performance-based contracts, 40, **41**, *42*, 228
    cycle-based versus, 30
peripheral zone, **86**, 228
personnel management
    call-out lists, 173
    crew communication training, 189, 197, 199, 200, 202, 205
    crew recognition, 190, 200
    grievances, 51
    labor unions, 50–52, 226
    leadership skills, 48–50, *49*
    performance evaluation, 49–50
    safety of crew, 5–6, 7, 15, 21
    safety of crew, storm response, 164, 183, 186–188, **187**, 189
    skills of, 47–*49*
    stakeholder interests, 199–200
phase (Ø), **134**, 228
phase-to-ground, 141, 228
phase-to-neutral, 141, 228
phase-to-phase, 141, 229
phone etiquette, 210, 211

photosystem I inhibitors, 113, *114*, 229
photosystem II inhibitors, 113, *114*, 229
pipe zone, **88**, 229
planning programs, 28–32
    quality assurance, 109
    scale of plan, 88–**90**, **89**
point sampling, 94, 229
poison, 111, 229
pollarding, **61**–62, 229
post-job review, 8, 190
potential energy, 13
practice storm drills, 173, **175**–176, 232
predictive maintenance, 29, 31, 229
pre-inspection contracts, 34, 40
pre-inspectors, 40, 43, 229
press releases, 207, 229
pre-staging, **181**–183, **182**, 229
preventive maintenance, 29, **59**, 229
    deferring versus, *72*, **74**, 221
primary line, 140, 141, **143**, 150, 229
    tree wire, **105**, *107*, 140, 234
primary voltage, 132, 150, 229
printed materials for outreach, 205, **206**
procurement departments, 44–45
professional development, 200
profit and loss reports, 39
profit centers, 36, 229
program budgeting, 38, 229
    budget justification, 199–200
program management
    budget justification, 199–200
    budgeting, 36–39, 219, 229
    circuit vs. grid work, 32
    contracting, 40–45, **41**, *42*, 44
    fleet management, 45–**47**
    inventory management, 117–**118**
    labor unions, 50–52
    personnel management, 47–50, *49*
    program planning, **28**–32
    project management, **32**–36, **34**, **35**
    quality assurance, 109
    revenue forecasting, 38
program planning, **28**–32
project management, **32**–36, **34**, **35**
protective grounds, 20

pruning cuts, **63–64**, **69**
    directional pruning, **70–71**, **73**, 221
pruning cycle, 29–30, 229
pruning machines, **76**, **99**, 229
pruning systems, 60–**62**, *61*, 96, 229
public relations
    angry public, 211–212
    communication importance, **197**–199, **198**
    community meetings, **207**
    customer relations versus, **204**
    definition, **204**, 229
    outreach methods, 205–**207**, **206**, 228
    public dissatisfaction, 198–199
    public perception, **198**
    public-friendly language, 211, 223
    storm brush disposal, 183–**184**
    storm response, 189
    support for utility pruning, 76, 196–197, **198**, 199, *201*, **210**
    talking with the public, **207**–212, *208*, **210**
    targeted messaging, 199–205, *201*, **202**, **203**
    work on display, **197**
punishment for incidents, 10, 21

Q

quadrat sampling, 94, 230
qualification, 200, 230
qualified line-clearance arborists, 230
    minimum approach distances, *15–16*, *17*
quality assurance, 109

R

radial distribution systems, 132, 230
reactive maintenance, 29, 31, 230
recordkeeping. *See* documentation
reduction cut, **63**–64, 230
regulations. *See* standards
regulatory agency messaging, 203
remote forest utility pruning, 75–77, **76**
removing trees, **75**
requests for proposals (RFPs), 45
resistance (electrical), **127**, 230
response growth, 67, 230
restoration of services, **164–165**, 230

restoration pruning, 188, 230
results-based management, 31, 230
returnable, reusable (R/R) containers, **115**, **116**, **118**, **119**, 230
reversibility, 111, 230
rights-of-way (ROW)
    definition, 230
    environmental protection, 91
    integrated vegetation management, 82, *84*
    pipe zone-border zone, **88**, 229
    remote forest utility pruning, 75–77, **76**
    scale of plan, 88–**90**, **89**
    site evaluation, 90–94, *93*, *93*
    targeted messaging, *201*–203, **202**
    tree risk assessment, 92–94, *93*, *93*
    utility pruning, **59**
    wire-border zone, 84–**87**, **85**, **86**, 235
right-to-work laws, 51–52
risk assessment matrix, 10–11, *12*
rosters, 183, 230
roundover, **62**, **70**, 230

S

safety. *See also* incidents; standards
    axioms of, 11
    behavior-based, 7, 8–13, **9**, 21
    culture of, 5, 7, 13, **21**–22, 230
    electrical safety precautions, 16, 19–21
    electrical safety standards, 14–*16*, *15*, *17*, 156, 227
    employer responsibilities, 5–6, 7, 15, 21
    job briefing, 8, 16, **17**, **18**, 225
    risk assessment matrix, 10–11, *12*
    risks, 4–5, 13–*14*
    safety committees, 6, 230
    stand-down, 186, 190, 230
    storm response crews, 164, 183, 186–188, **187**, 189
    training, 7, 11, **21**
    worker responsibilities, 7, 8, 11, **21**
    worker rights, 6–7
    worker safety tools, 8
Saffir-Simpson scale, *168*, 230
SAIDI (System Average Interruption Duration Index), 151–152, 231

SAIFI (System Average Interruption Frequency Index), 151–152, 231
sampling methods, 94
SCADA (supervisory control and data acquisition), 129, 233
scale of plan, 88–**90**, *89*
secondary lines, 141, 150, 231
secondary voltage, *129*, **132**, 150, 231
seed bank, 105, 231
selective foliar applications, *102*, 103, 231
selective herbicides, 110, 231
self-checking S.T.A.R., 8
service containers, 115, 231
service drop, 141, **142**, 231
service interruption, 58, 231. *See also* interruptions
service line, 141, **142**, 231
set objectives, **83**, 231
shear, 231
shearing, 62, 231
Shigo, Alex, **73**
short circuits, 128, 152–**153**, 231
side-pruning, *102*, 103, 231
simple cash accounting, 39, 231
sine wave, 133–**134**, 231
single point grounding, 20
site conditions, 90–94, **93**, *93*, 231
sleet, 170, 231
small-scale responses, 180, 231
SMART goals, 6, 50, *84*, 231
snow, 171, **172**
social media
    crews at storm response, 189
    customer dissatisfaction, 198–199, 202
    definition, 231
    outreach methods, 205, 206–207
soil drench, **100**, 232
soil injection, 100, **101**, 232
solar generation, **137–138**, 232
span guy, 141, 232
spark-over, 95, 232
specifications, 28–29, 232
    lump-sum contracts, 43, *44*
    unit price contracts, 42–43
staging areas, 175, 176, 180, **181**, 232

stakeholders
    angry stakeholders, 211–212
    communication importance, **197**–199, **198**
    community meetings, **207**
    definition, 232
    external stakeholders, 200–**203**, *201*, **202**
    internal stakeholders, 199–200
    outreach methods, 205–**207**, **206**, 228
    stakeholder dissatisfaction, 198–199
    stakeholder perception, **198**
    stakeholder-friendly language, 211, 223
    support for utility pruning, 76, 196–197, **198**, 199, *201*, **210**
    talking with, **207**–212, *208*, **210**
    targeted messaging, 199–205, *201*, **202**, **203**
    work on display, **197**
standards. *See also* ANSI A300; ANSI Z133; FAC-003
    American National Standards, 19
    clearance regulations, 154–158, *157*
    electrical safety, 14–*16*, *15*, 156, 227
    fire rules and regulations, 157–158
    National Electrical Safety Code® (NESC), 156, 227
    New Zealand regulations, *157*
    regulatory agency messaging, 203
    specifications based on, 29
    transmission vegetation management, 83
    utility pruning, 60
start-to-start dependency, 33
state of emergency, 185–186, 232
stator, **133**, 134, 232
steam turbine plants, **133**, 134–135
step potential, 16, 20, 232
step-down transformers, 148, 150, 232
step-up transformers, 148, 150, 232
storm center, **173**, 232
storm drills, 173, **175**–176, 232
storm preparation
    communications check, 173
    crew equipment, 185
    deployment planning, 172, **175**–177, 183
    Incident Command System, 172–173
    monitoring conditions, 176–177
    mutual assistance, 177
    planning ahead of time, 172

pre-staging, **181**–183, **182**
storm drills, 173, **175**–176
supplier arrangements, 176, 188
storm recovery
    brush disposal, 183–**184**, 186
    crew recognition, 190
    crew release, 189
    definition, 232
    lessons learned, 190
    reimbursement of contractors, 188
    saving damaged trees, 188
    steps involved, **164–165**
storm response
    brush disposal, 183–**184**, 186
    crew accommodations, 180–**181**, **182**, 184–185, 186
    crew equipment, 183, 185
    crew release, 189
    crews, 179–180, 183–185, **184**
    definition, 232
    deployment planning, 172, **175**–177, 183
    documentation, 188
    emergency declarations, 185–186
    emergency operations center, **173**
    Incident Command System, 165, 172–173
    large-scale responses, 180–185, **181**, **182**, **184**, 226
    loads on trees, 165–**166**, 170–**171**, 172, 226
    media relations, 189
    pre-staging, **181**–183, **182**
    safety concerns, 183, 186–188, **187**
    saving damaged trees, 188
    small-scale responses, 180, 231
    steps involved, **164–165**
storm surge, 168, 232
storm types
    cyclones, 166–*168*, **167**, **174**–175, 221, 224, 234
    scales ranking, 167–*168*, *170*
    thunderstorms, 168–**170**, **169**, *170*, 233
    tornados, 169–**170**, **171**, 233
    track forecast cone, **178–179**
    warnings, 166
    watches, 166
    winds, locally high, 172
    winter storms, 170–**171**, **172**
straight-line depreciation, 37
straight-line winds, 169, 232

strategic vs. tactical planning, 28
stream protections, 91
stress, 71, 165–**166**, 232
strike zone, 68, 232
struck-bys, 4, 13–14
stump applications, *102*, 232
subordination, 65–66, 232
substations, **129**, **130**, **131**, 132, 232
subtransmission lines, **129**, **130**–131, 232
supervisory control and data acquisition (SCADA), 129, 233
supply containers, **115**, **116**, 117, **118**, **119**, 233
switches on lines, 143, **144**
switchyards, 130, 233
SWOT analysis, **28**, 233
synthetic auxins, 110, 113, *114*, 233
system reliability indices, 151–152

T

tactical vs. strategic planning, 28
taper, 166, 233
taps, **131**, 145, 233
targeted messaging, 199–205, *201*, **202**, **203**, 227, 233
telephone lines
    joint facilities, 141, **143**
    storm response, 182, 184
teratogenicity, 111, 233
terminal bud, 233
text messaging for outreach, 206
thunderstorms, 168–**170**, *169*, *170*, 233
time and material contracts, 40–**41**, *42*, 233
    lump sum versus, 43
tolerance levels, 82, 94–95, 233
topiary, 62, 233
topping, **62**, **70**, 233
    epicormic shoots, **66**, **70**, 222
tornados, 169–**170**, **171**, 233
    cyclones versus, **167**, 169
    Enhanced Fujita Scale, *170*, 222
touch potential, 16, 20, 233
toxicity, *110*–112, 233
track forecast cone, **178–179**, 233
training
    crew communication skills, 189, 197, 199, 200, 202, 205

professional development, 200
safety, 7, 11, **21**
talking with customers, **207**–212, *208*, **210**
turnover rates and, 200
transformers, **132**, 141, 148, **150**–151, 233
transient faults, 128, 233
transmission interconnect, 129–130, 233
transmission blackouts, 153–154, **155**
transmission of electricity, **129–130**
circuit protection, 144–145
equipment, *148*–**151**, **149**, **150**
transmission blackouts, 153–154, **155**
transmission lines, **129**, *129*, 234
transmission towers, **139**
voltage classifications, *129*
trauma, 13–14
tree growth regulators (TGRs), 72, 99–**101**, **100**, 112, 234
tree on line, 19, **187**
tree-caused interruptions, 151–**154**, **153**
tree protection zones (TPZs), 106, *108*, 234
tree risk assessment
definition, 234
integrated vegetation management, 92–94, **93**, *93*
prune or remove, **75**
risk assessment, 68
site evaluation, 90–94, **93**, *93*
Tree Risk Assessment Qualification program, 92, **93**, *93*
workload assessments, 91
tree wire, **105**, *107*, 140, 234
triple constraint, **32**, 234
triplex wires, 141, **142**, 234
tropical cyclones, **167**–*168*, **174**–175, 234
tracking, **178–179**
trunk injection, 100, **101**, 234
turn-key contract, 182, 188, 234
turns ratio, 150, 234
twisters. *See* tornados
two-minute rule, 8
typhoons, 166–*168*, **167**, **174**–175, 221, 224, 234

U

underground construction, 105–106, *107*, **143**, 234
unions (labor), 50–52, 173, 226

unit price contracts, 40, **41**–43, *42*, 234
unsafe acts, 4, 234
unsafe working conditions, 4, 234
urban forest
balanced with utility pruning, ix, 58
definition, 234
greenspace inventories, 88–89
prune or remove, **75**
stakeholder targeted messaging, *201*–**202**
workload assessments, 91–92
Urban-Wildland Interface Code (2006), 157
utilities as stakeholders, 199, 200
utility facilities
deferring maintenance, *72*, **74**, 221
definition, 234
overhanging branches, 71, **72**
utility pruning, 58–**59**
utility forest, 58, 234
utility pruning
contributions of Alex Shigo, **73**
deferring maintenance, *72*, **74**, 221
definition, 234
directional, **70**–**71**, **73**, 221
easements, 59–**60**, 222
generating support for, 76, 196–197, **198**, 199, *201*, **210**
integrated vegetation management and, 96
objectives, **66**–**71**, **69**, **70**
palm pruning, **74**–75
prune or remove, **75**
pruning cuts, **63**–**64**, **69**
pruning intervals, 71–74, *72*
pruning systems, 60–**62**, **61**, 96, 229
purpose of, 58–60, **59**
remote forests, 75–77, **76**
restoration pruning, 188, 230
standards and best practices, 60
structural, 64–**66**, **65**
tree response to, **64**
urban forest balance, ix, 58
winter storms, **171**, **175**
work on display, **197**
utility regulatory agency messaging, 203
utility service
"continuous and dependable," vii
definition, 234

utility pruning and, 58, **66–67**
utility vegetation manager, 235. *See also* program management

V

variable costs, 36
vegetation control methods. *See* control methods
vegetation management. *See* integrated vegetation management; program management
vehicle monitoring system (VMS), **47**, 235
vehicles in fleet, 45–**47**
vehicles under downed lines, 20
vigor, 62, 64; 235
volt (V), 126, **127**, *129*, 235
voltage gradient, 152–153, 235
voltage ratio, 150, 235
voltage regulators, **151**, 235

W

warning (storm), 166, 235
watch (storm), 166, 235
water head, 135, 235
watt (W), 126–**127**, 150–151, 235
websites for outreach, 206
Weingarten rights, 50–51, 235
wetland protections, 91
whorls, 71, 235
willful rule violation, 4

wind turbines, 135–137, **136**
windrow, 108, 235
winds, locally high, 172
winter storms, 170–**171**, **172**
wire zone, 84–**85**, 103
wire-border zone, 84–**87**, **85**, **86**, 235
    cover-type conversion, 104–105
    cut-stubble herbicide applications, 103
    pipe zone, **88**, 229
wires, 140–**143**, **142**, 220, 222
work breakdown structure, 33, 235
work planners, 34–36, **35**
work scheduling, 29–30
workload assessment, 91–92, 235
wound treatment, **64**, 235
Wye construction
    delta construction versus, 141, 221
    distribution transformer, **132**
    explanation of, 141, 235
    single point grounding, 20

Y

"the yellow book," **73**

Z

Z133. *See* ANSI Z133
zero-based budgeting, 38